ROUTLEDGE LIBRARY EDITIONS: ENERGY RESOURCES

Volume 6

THE NUCLEAR POWER DEBATE

THE NUCLEAR POWER DEBATE

A Guide to the Literature

JERRY W. MANSFIELD

Routledge
Taylor & Francis Group

LONDON AND NEW YORK

First published in 1984 by Garland

This edition first published in 2019
by Routledge
2 Park Square, Milton Park, Abingdon, Oxon OX14 4RN

and by Routledge
52 Vanderbilt Avenue, New York, NY 10017

Routledge is an imprint of the Taylor & Francis Group, an informa business

British Library Cataloguing in Publication Data
A catalogue record for this book is available from the British Library

ISBN: 978-0-367-23168-2 (Set)
ISBN: 978-0-429-27857-0 (Set) (ebk)
ISBN: 978-0-367-23127-9 (Volume 6) (hbk)
ISBN: 978-0-367-23130-9 (Volume 6) (pbk)
ISBN: 978-0-429-27839-6 (Volume 6) (ebk)

Publisher's Note
The publisher has gone to great lengths to ensure the quality of this reprint but points out that some imperfections in the original copies may be apparent.

Disclaimer
The publisher has made every effort to trace copyright holders and would welcome correspondence from those they have been unable to trace.

THE NUCLEAR POWER DEBATE
A Guide to the Literature

Jerry W. Mansfield

GARLAND PUBLISHING, INC. • NEW YORK & LONDON
1984

Library of Congress Cataloging in Publication Data

Mansfield, Jerry W.
The nuclear power debate.

Includes index.
1. Atomic power—Bibliography. I. Title.
Z5160.M36 1984   016.62148   83-48255
[TK9145]
ISBN 0-8240-9102-7 (alk. paper)

Cover design by Larry Walczak

Printed on acid-free, 250-year-life paper
Manufactured in the United States of America

To My Parents
William and Beverley

CONTENTS

Preface

All over this world there are people who
watch complacently as citizens in other parts of
the United States and abroad rise up against
nuclear power or as others speak in its favor.
Often the two views are contradictory and the
general public is confused about the real issues.
One day there is talk of a nuclear power plant to
be built near your home or workplace and suddenly,
you have neighbors and friends who have never
publicly supported anything or who have never
before protested against an issue poised for action.
People on both sides of the nuclear power debate
will begin to seek information to support their
views and to further educate themselves on the
crucial issues.

This annotated bibliography will prove to be
a starting point to that education and awareness
of the issues. The nuclear power debate is a com-
plicated one and it is hoped that through the
reading of the books in this list, most of which
should be available in medium to large public
libraries, citizens will be far more aware and able
to take a firm stand in their belief, for whatever
side of the question they support.

Since the 1979 Three Mile Island incident the
marketplace has been flooded with books on nuclear
power and alternative energy sources. Many of
these were written overnight to satisfy the public
demand for more information on radiation, wastes,
nuclear hazards, transportation and other related

topics. The thirst for nuclear energy information was insatiable and books were followed by
magazine articles and newspaper stories. Some
books were written by knowledgeable experts while
others were haphazardly put together merely to
capitalize on this nuclear information craze.

Thousands of government and industrial technical reports are available on nuclear energy.
Even more journal articles have been written in the
popular and scientific literature. It would be a
massive undertaking to identify and read all that
has been written since 1975 on the nuclear power
question. This bibliography is, therefore, limited
to books of a non-technical, general interest
nature written primarily in the last decade. No
attempt was made to exclude titles or works by
certain authors. Rather, the intention is to
present a cross-section of reading materials in a
balanced format.

In the last few years more books have been
written *against* nuclear power than *for* this energy
source. One may speculate that those people who
are against any movement are more vocal than the
supporters, that it is easier to publish and market
an anti-nuclear book, or that nuclear activists
must try twice as hard to get their point across.
There are also well-financed and highly visible
lobbying groups against nuclear power, but smaller,
quieter ones for nuclear growth. As one of the
authors in this bibliography suggests, nuclear
advocates seem less vocal because they are lost in
the din of the ignorant and, in time, even nuclear
activists will cross over to the other side out of
necessity.

The intent of this research guide is not to
say that one side is right and another is wrong.
Rather, it is to provide a selected sampling of
quality, non-technical books from all viewpoints
in the nuclear power debate.

The book is arranged in three sections.
First there are the books which have been deter-
mined to be PRO-NUCLEAR, followed by those which
are decidedly ANTI-NUCLEAR. The last section
consists of some books which are NEUTRAL treatments

of the debate. In all sections the books are
arranged alphabetically by the authors' last
names unless the book is a collection of essays or
has no author, in which case it is listed by the
title.

JWM

Pro-Nuclear

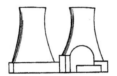

1.   Beckman, Petr. THE HEALTH HAZARDS OF <u>NOT</u>
     GOING NUCLEAR. Boulder, CO: Golem, 1976.

     Starting off this section is a powerful
pro-nuclear book with a unique angle. In
less than polite tones Beckman critizes
Ralph Nader's ignorance of nuclear safety
and states that the nuclear debate, if there
is anything to debate, is replete with myths,
distortions, and falsehoods. We are informed
that an atomic explosion in a nuclear plant
is physically impossible due to the type of
uranium used as a power plant fuel. Several
times the author makes it clear to the reader
that "the point of this book is not to argue
how dangerous fossils are, but that nuclear
power is safer." The book is full of anger
toward all the unqualified people who speak
out against nuclear power. Examples include
Nobel Prize winners who are distinguished in
fields unconnected with nuclear power but
who speak against it as if the Nobel Laureate
title makes them instant experts in all
areas of science. Further verbal lashings
are given to the media who have a bias
toward anti-nuclear causes, the misleading

statistics which are used, and of course,
the environmental and political action
groups which spring up like mushrooms at the
mere mention of the word "nuclear."

2.   Bromberg, Joan Lisa. FUSION: SCIENCE,
       POLITICS, AND THE INVENTION OF A NEW
       ENERGY SOURCE. Cambridge, MA: MIT Press,
       1982.

       From a historical viewpoint Dr. Bromberg
provides a very readable, quasi-scientific
account of controlled thermonuclear research
(CTR). For over thirty years researchers
have been trying to demonstrate that nuclear
fusion reactions between light elements can
provide a new energy source. Concentrating
on the political and organizational aspects
of the Magnetic Fusion Program in the U.S.,
we learn of the early security clamps placed
on the research centers of Los Alamos, Oak
Ridge, and Princeton. The classified aspect
of CTR was weakened in 1958 in time for the
Second Geneva Conference where all countries
with fusion programs displayed their pro-
gress. The American exhibit, a flashy
mechanical display, was the Sherwood Specta-
cular which attracted many visitors. Project Sherwood was the name given to the Los
Alamos project. The latter portions of
Fusion are devoted to the effects which the
many changes in leadership, industrial and
political, have had on the various research
programs. Albeit this is not a book direct-
ly concerned with the nuclear controversy
of the 1980s, Bromberg's account may be the
first official history of any nation's
fusion research and for that it is a noteworthy document. The book was commissioned
by the Office of Magnetic Fusion Energy of
the U.S. Department of Energy.

3.   Bupp, Irvin C., and Jean-Claude Derian.
      LIGHT WATER: HOW THE NUCLEAR DREAM
      DISSOLVED. New York: Basic, 1978.

         After World War II the hopes for nuclear
      energy as a power source were almost real-
      ized and became a greater possibility after
      the 1973 OPEC oil embargo. This book is an
      interpretation of how the most promising
      reactor technology, light water, seemed to be
      the answer to the nuclear energy need. The
      downfall came in the manner in which the U.S.
      managed and presented the need for nuclear
      power to the citizens. Distrust was the
      result and these emotions are still the
      driving force behind the nuclear safety
      controversy. The involvement of the corpor-
      ations, politicians, and the citizens groups
      for and against nuclear energy are clearly
      charted in this historical assessment of
      what is, according to the authors, a "dream
      dissolved."

4.   Cameron, I.R.   NUCLEAR FISSION REACTORS.
      New York: Plenum, 1982.

         This is a sound and thorough introduc-
      tory text on fission technology. A diverse
      presentation of scientific and technical
      ideas is provided in a light, readable and
      clear manner. Cameron discusses the physics
      of atoms as it relates to nuclear reaction
      theory and follows with explanations of
      nuclear fuel, radiation damage, and basic
      heat transfer. Four chapters are devoted
      to the main reactor types including light-
      water, fast-breeder reactors, heavy-water,
      and gas-cooled systems. The author rounds
      out the book with a very long chapter on
      the safety and environmental effects of
      nuclear power including a brief account of
      Three Mile Island.

5.   Cantelon, Philip L., and Robert C. Williams.
      CRISIS CONTAINED: THE DEPARTMENT OF
      ENERGY AT THREE MILE ISLAND. Carbondale,
      IL: Southern Illinois University Press,
      1982.

          The Department of Energy commissioned
      the authors to chronicle the months of acti-
      vity following the 1979 incident at Three
      Mile Island. DOE's involvement included
      monitoring, analyzing, and coordinating the
      radiation data from TMI. The conclusion
      reached was that human errors were to blame,
      not technological failures. Two useful
      appendices include a roster of all personnel
      investigating the TMI aftermath and a
      synoptic chronology of events.

6.   Clark, Ronald W. THE GREATEST POWER ON
      EARTH: THE INTERNATIONAL RACE FOR SUPRE-
      MACY. New York: Harper, 1981.
      (Published in England as *The Greatest
      Power on Earth: The Story of Nuclear
      Fission.* London: Sidgwick & Jackson,
      1980.)

          The major portion of this work covers
      the history of the discovery of atomic
      fission in 1938 through the wartime efforts
      to produce nuclear weapons in Britain, the
      U.S., Germany, and Russia. Careful reading
      of the book will make clear the distinction
      between fission and fusion energy. Clark
      provides a brief account of the development
      of the hydrogen bomb, the attempts to uti-
      lize fission power as an energy source, and
      he touches upon the growing public awareness
      of the radiation perils associated with
      peaceful and military uses of nuclear power.
      In his concluding paragraph he says, "It
      would be tragic if a limited nuclear
      exchange, halted after each side had exper-
      ienced the agony of one city put to the bomb,
      were necessary to bring men to their senses
      and to realize , at last, that there are no
      victors in nuclear war."

7.   Cochran, Thomas B. THE LIQUID METAL FAST
      BREEDER REACTOR. Baltimore: Johns
      Hopkins University, 1974.

         The Liquid Metal Fast Breeder Reactor
      (LMFBR) has been subject to considerable
      scrutiny and criticism in the U.S. despite
      its success and expansion in the UK, France,
      Germany, Italy, Japan, and the Soviet Union.
      Mr. Cochran takes a serious look at the
      costs and the environmental arguments which
      favor the LMFBR use in the U.S. Relying
      heavily upon AEC reports he assesses fuel
      cycle costs, performance, and even uranium
      supply and demand. The author tends to be
      pessimistic about the future of LMFBRs and
      instead advocates thermal reactors and a
      continued search for responsible energy
      alternatives. A critical analysis is made
      between the LMFBR and high-water reactors
      and Cochran suggests that perhaps the U.S.
      should wait to see what new information and
      technology is available by 1986 to ensure
      that the decision made is one which will
      last into the next century. The problems of
      waste storage, safe fuel transportation,
      and reactor safety are all touched upon in
      this critical look at the liquid metal fast
      breeder reactor.

8.   Cohen, Bernard L. NUCLEAR SCIENCE AND
      SOCIETY. Garden City, NY: Anchor, 1974.

         A well-read individual of nuclear topics
      would understand and appreciate the correla-
      tions made by Dr. Cohen between nuclear
      research and society's needs. In this work
      we are provided with some benefits of
      radiation not generally known, such as its
      ability to retard food spoilage, control

insects, and aid in genetic crop research.
Other uses include the desalting of sea-
water as well as peacetime ship and rocket
propulsion. Along with a historical view of
society's use, value, and perceptions of
energy, the author provides concise and
graphic descriptions of how fission reactors
and fast breeder reactors work. Finally,
brief coverage is given for the peaceful
uses of nuclear explosives, nuclear warfare,
and nuclear weaponry.

9.   Cottrell, Alan, Sir. HOW SAFE IS NUCLEAR
      ENERGY? Exeter, NH: Heinemann, 1981.

      This is a very factual book about how
reactors produce energy, how wastes are
released, the effects of varying levels of
radiation, and the question of safety. The
author is admittedly pro-nuclear but here he
presents all of the facts and allows the
reader to decide for himself. If nuclear
power is unsafe, then we should reject it.
If it is safe or can be made safer, then
let's get rid of the fear and get on with it.
Should this world go forth with increased
production and dependence on nuclear power,
then we owe future generations a safe and
reliable energy system. Sir Alan's comments
are sure to provoke both the anti-nukes with
his strong support for nuclear power and the
nuclear establishment with his concerns about
the fracture of pressure vessels.

10.   Garvey, Gerald. NUCLEAR POWER AND SOCIAL
          PLANNING. Lexington, MA: Lexington,
          1977.

          Believing that America's commitment to
       nuclear power is irreversible, this Princeton
       professor of politics advocates that we learn
       how best to control nuclear development. The
       technical, economic, and socio-political
       factors affecting the question of control are
       carefully examined. Garvey discusses the
       need to switch from fission to fusion energy
       and asks if power parks, or nuplexes
       (nuclear powered agro-chemico-industrial
       complexes) as they are sometimes called, are
       a solution. Many useful charts and diagrams
       help the reader to understand the issues and,
       perhaps for the undecided, aid in making up
       their own mind in this debate.

11.   Glasstone, Samuel, and Walter H. Jordan.
          NUCLEAR POWER AND ITS ENVIRONMENTAL
          EFFECTS. New York: American Nuclear
          Society, 1980.

          Contrary to what anti-nuclear critics
       say, Drs. Glasstone and Jordan state that it
       is possible for nuclear power to account for
       one-fifth of the U.S. electricity generated
       by 1990. They do realize, however, that a
       successful nuclear program in the U.S. will
       depend on the public's perception of its
       environmental and safety aspects. Knowing
       this, the authors clearly describe the
       environmental effects of nuclear power and
       explain the safety measures taken to ensure
       the safe operation of power plants. The
       needless confusion which exists on this
       topic is cleared by their numerous illus-
       trations, examples, and their combined years
       of first-hand experience.

12.   Grenon, M. THE NUCLEAR APPLE AND THE SOLAR
          ORANGE: ALTERNATIVES IN WORLD ENERGY.
          Elmsford, NY: Pergamon, 1981.

          Affiliated with the International Insti-
       tute for Applied Systems Analysis in
       Austria, the author utilizes his "watchdog"
       position to assess the world energy situa-
       tion. His non-technical discussion includes
       heavy emphasis on nuclear and solar energy
       as the primary sources of world energy.
       This overview of the energy dilemma also
       includes the merits of geothermal energy and
       coal. Not limited to U.S. energy problems,
       this work covers the global question and the
       need for alternative energy sources.

13.   The Harvard Nuclear Study Group. LIVING
          WITH NUCLEAR WEAPONS. Cambridge, MA:
          Harvard University Press, 1983.

          The President of Harvard University,
       Derek Bok, urged the University to take a
       stand on the issue of nuclear weapons and
       Living With Nuclear Weapons is their state-
       ment. The work is a valuable history of how
       the present nuclear predicament developed
       and serves as a guide to policy options for
       coping with the situation. One of the book's
       major points is that the public spends far
       too much time worrying about all of the
       fringe areas of nuclear power such as
       mechanical/human errors and sabotage rather
       than the real dangers of war in Third World
       countries who are newcomers to the nuclear
       weapon arena. Numerous arguments are made
       for ICBMs, MX missiles, and more non-contro-
       versial weaponry. Although this work does
       not inform readers how to get rid of the
       nuclear war threat, it will nevertheless be
       a landmark, controversial book on the
       nuclear power debate.

14.   Hertsgaard, Mark. NUCLEAR INC.: THE MEN AND
       MONEY BEHIND NUCLEAR ENERGY. New York:
       Pantheon, 1983.

       This work is a history of the atomic-
       industrial complex and the people and corpor-
       ations which make it function. The author
       interviewed executives from the major firms
       which hope to sell reactors, construct the
       power plants, and provide the nuclear power
       generated electricity to make for a nuclear
       powered America. Briefly mentioned is the
       relationship which these dozen or so firms
       have with big business and government. A
       few of the executives reveal the "Atomic
       Brotherhood's" (that's what they call them-
       selves) plans and strategies for making
       America dependent on safe and efficient
       nuclear power.

15.   Hoyle, Fred, Sir, and Geoffrey Hoyle.
       COMMONSENSE IN NUCLEAR ENERGY. New York:
       Freeman, 1980.

       Many myths surround the nuclear power
       industry and this father and son team try to
       dispel some of them. Arguing that fossil
       fuels will run out someday, they develop
       the case for nuclear power as a clean,
       viable, and safe alternative. The authors,
       both renowned scientists and conservation-
       ists, point out that technologists and
       environmentalists can and must learn to live
       harmoniously. This is a well-written and
       easy-to-read pro-nuclear book.

16.   Hoyle, Fred, *Sir*.   ENERGY OR EXTINCTION?
       THE CASE FOR NUCLEAR ENERGY. London:
       Heinemann, 1979.

            Forever the crusader for nuclear energy,
       Hoyle illustrates that we will someday run
       out of oil and that we really have no prac-
       tical alternative fuel. Although the public
       is favorable to solar energy and the media
       remains optimistic, the technology to har-
       ness this and other alternative energy
       sources on a large scale just does not exist.
       It is safe, reliable, and power can be cre-
       ated without fast-breeder reactors like the
       CANDU (Canadian-Deuterium-Uranium) type
       reactor. In his usual controversial fashion
       the author points out that some of the anti-
       nuclear campaigns are politically inspired
       as a means to weaken the West. Written in
       non-scientific terms, this book serves as an
       important introduction to the nuclear energy
       controversy.

17.   Imai, Ryukichi, and Henry S. Rowen.
       NUCLEAR ENERGY AND NUCLEAR PROLIFERATION:
       JAPANESE AND AMERICAN VIEWS. Boulder,
       CO: Westview, 1980.

            This elementary foreign policy book
       treats the sensitive international issue of
       nuclear nonproliferation. When President
       Carter came into office in 1977 he promised
       better U.S.-Japanese relations and a closely
       guarded nonproliferation policy. Carter
       announced that the U.S. was going to defer
       the use of plutonium in existing reactors
       and that a restructuring of breeder reactor
       research and development was to occur which
       would stress safety rather than early
       commercialization. This was in direct con-
       flict with an existing agreement with Japan
       which pledged to forge ahead. The authors
       of this work were involved in behind-the-
       scenes activity to arrange for a solution
       amenable to both countries. Mr. Imai, as

general manager of engineering of the Japan
Atomic Power Company and as Japanese mini-
sterial advisor, and Mr. Rowen, a Stanford
Professor, former Rand Corporation president,
former assistant director of the U.S. Bureau
of the Budget, and deputy assistant secretary
of defense, spent years working out the
details and acting as go-betweens on this
vital issue. This is their side of the
story.

18.   Knief, Ronald Allen. NUCLEAR ENERGY TECH-
       NOLOGY: THEORY AND PRACTICE OF COMMERCIAL
       NUCLEAR POWER. Washington, DC:
       Hemisphere, 1981.

       Knief's book is intended to be an author-
atative reference book on nuclear energy
technology for the scientist as well as the
student. Theory and the practical aspects
of nuclear technology are provided to assist
in the understanding of current practices,
reactors, and future prospects. A review
of fuel cycles, reactor design, and reactor
safety fundamentals is provided. The status
of nuclear energy worldwide is presented by
this former manager of training at Three
Mile Island. Extensive bibliographies
follow each chapter.

19.   Lapp, Ralph Eugene. NADER'S NUCLEAR ISSUES.
       Greenwich, CT: Fact Systems, 1975.

       Dr. Lapp declares war on Ralph Nader's
charge that "nuclear fission power is
unsafe, unnecessary, and unreliable" in this
battle of words. The format for the majority of the book is that of actual Nader
comments followed by a rebuttal from the
author. Numerous appendices are included
which contain endorsements for nuclear power
from scientists, congressmen, and Nobel
Prize winners.

20.   Libby, Leona Marshall. THE URANIUM PEOPLE.
      New York: Crane, Russak/Scribners, 1979.

          This is a historical book of the atomic
      stars who were involved in the early years
      of discovery and development of atomic
      energy. Dr. Libby was the youngest and the
      only female member of the group which built
      the first reactor. As a physicist she
      worked in the company of Robert Oppenheimer,
      Glenn Seaborg, and other Nobel Laureates-to-
      be. A special insight into the problems of
      physics and chemistry which the scientists
      encountered is provided. However, more than
      telling what was scientifically interesting
      about the uranium-plutonium project, Libby
      has a flair for telling what was human and
      fascinating about the people involved. She
      often reveals the human thoughts of the
      scientists as they worked out scientific and
      moral problems. This documentary with black
      and white photos ends with the establishment
      of the Atomic Energy Commission, the Russian
      A-bomb, and the H-bomb controversy.

21.   Lilienthal, David E. ATOMIC ENERGY: A NEW
      START. New York: Harper, 1980.

          This is the premier history of nuclear
      energy research and development in the U.S.
      written by a man who once headed the
      Tennessee Valley Authority and who was the
      first chairman of the Atomic Energy Commi-
      ssion. As an advocate of nuclear responsi-
      bility, Lilienthal argues for safer, peaceful
      uses of nuclear power and the elimination of
      nuclear weapons. Some predictions of the
      future of nuclear energy are also ventured.

22. Longstaff, Malcolm. UNLOCKING THE ATOM: A
     HUNDRED YEARS OF NUCLEAR ENERGY.
     London: Muller, 1980.

        Most of the recent books on the develop-
     ment of nuclear energy dwell on the political
     and moral arguments yet few can succeed in
     covering the technological aspects in a
     manner which is appealing to the general
     reader like *Unlocking the Atom*. The book
     begins with the story of the discovery of
     uranium, radioactivity, and fission and
     discusses the work leading to the bombs of
     Nagasaki and Hiroshima. Full chapters each
     are devoted to safety, energy, the place of
     nuclear power today, and other uses of
     nuclear power. Well-chosen photographs
     complement the text. This easy-to-read
     history was written by a twenty-four year
     veteran of the United Kingdom Atomic Energy
     Authority.

23. McCracken, Samuel. THE WAR AGAINST THE ATOM.
     New York: Basic, 1982.

        This is another non-technical, historical
     review of the atomic industry. McCracken
     examines individuals and organizations
     behind the anti-nuclear movement in order to
     more fully understand the roots of resistance.
     In his conclusion the author attacks the
     views of five well-known anti-nuclear
     activists. This critical last chapter alone
     makes the reading of the book worthwhile.
     This is very much a one-sided, pro-nuclear
     essay describing the battle against the
     nuclear industry.

24. Maddox, John. BEYOND THE ENERGY CRISIS.
New York: McGraw-Hill, 1975.

In response to the OPEC oil crisis, this
British physicist pleads for reduced oil
consumption and the development of alternative sources of fuel. The favored source
for the author is nuclear generated power.
This is a factual account of the economic,
political, and technical difficulties of
converting nuclear power into a viable source
of energy. This would be a more valuable
book if it had tables and graphs illustrating
the author's points, but nevertheless, it is
a good synopsis of the energy crisis and has
some positive solutions.

25. Mann, Martin. PEACETIME USES OF ATOMIC
ENERGY. Rev ed. New York: Viking,
1961.

Here is an excellent book extolling the
usefulness of atomic energy in everyday life.
Mann illustrates how our quality of life
has been enhanced by atomic energy, how it
has actually protected out health, and how
it aids us in answering the how and the why
of the universe. Numerous illustrations
help the reader to understand the positive
impact of atomic power. This should be
read in conjunction with the Cohen book,
Nuclear Science and Society. (Item number 8)

26.   THE NUCLEAR POWER CONTROVERSY. Englewood
      Cliffs, NJ: Prentice-Hall, 1976.

      The American Assembly, an affiliate of
      Columbia University, has sponsored this
      series of papers on the nuclear debate.
      This book contains five general essays by
      pro-nuclear specialists and one by a nuclear
      critic as a balance. Topics covered include
      reactor safety, including waste storage, the
      economics of nuclear power, a brief history
      of the ups and downs in the twenty years of
      nuclear power generation, the nuclear
      regulatory process, the international aspects
      of the controversy as well as nuclear pro-
      liferation, and the anti-nuclear power essay
      which asks, "How much is too much?" This
      collection of papers is intended to start a
      dialogue on the subject of nuclear power
      and to encourage the general public to
      continue questioning the technologies and
      policies of others which pervade their own
      lives.

27.   Schmidt, Fred H., and David Bodansky. THE
      FIGHT OVER NUCLEAR POWER: THE ENERGY
      CONTROVERSY. San Francisco: Albion,
      1976. Foreword by Hans Bethe.

      Realizing that the U.S. needs energy,
      that fossil fuels are limited, that uranium,
      thorium, and breeder reactors can satisfy
      the energy needs of the world for 100,000
      years, and that nuclear power is safe, the
      authors push for the urgent, responsible
      development of nuclear power. The energy
      alternatives heralded by other writers are
      viewed as insufficient for large-scale
      energy production. However, researchers
      must continue to search for scientific break-
      throughs in the area of alternative solu-
      tions to the dwindling fossil fuel supply.
      The problems of nuclear power are not over-
      looked and some solutions are offered which
      are thought provoking.

28.   Seaborg, Glenn T., and William R. Corliss.
       MAN AND ATOM: BUILDING A NEW WORLD
       THROUGH TECHNOLOGY. New York: Dutton,
       1971.

       The apprehensions and concerns of anti-
       nuclear individuals are understood and an
       attempt is made to allay their fears by
       showing the peaceful applications of the
       atom. Nuclear power technology and research
       has given the world numerous medical,
       household, and outer space uses. Even
       greater than these are the future prospects
       for benefit to mankind in the areas of agri-
       cultural nuplexes and underwater exploration
       and habitation. With no intention of side-
       stepping the issue of nuclear energy in
       general, the authors contend that the
       scientific community will work out solutions
       to these problems and that right now our
       need is for a viable solution to the fossil
       fuel energy drain. Directed to the intelli-
       gent non-technical reader, this book has
       many useful photos and charts to reinforce
       its arguments.

29.   Smart, Ian. WORLD NUCLEAR ENERGY: TOWARD A
       BARGAIN OF CONFIDENCE. Baltimore:
       Johns Hopkins University Press, 1982.

       In 1977 and with a membership of just
       twenty people the International Consultative
       Group on Nuclear Energy (ICGNE) was formed
       to examine the uncertainties of the nuclear
       industry. *World Nuclear Energy* brings
       together eight of ICGNE's position papers
       written between 1978 and 1980 which cover
       international nuclear relationships, the
       history of world atomic cooperation, the
       economic and technical constraints of
       reactors, and nuclear plant construction.
       ICGNE supports an expanded program of inter-
       national cooperation in fast reactor develop-
       ment. The authors of these papers also

argue for an international program of
waste management and a demonstration of the
commercial feasibility of spent fuel repro-
cessing. Although they advocate a total
international cooperation and idea sharing
in nuclear power, they also realize that the
major roadblock to this is the fear of
weapons proliferation. The essays in this
book provide some valuable nuclear concerns,
some of which we will still be discussing
twenty years from now.

30.   Williams, Roger. THE NUCLEAR POWER DECI-
      SIONS: BRITISH POLICIES, 1953-78.
      London: Croom Helm, 1980.

      Twenty-five years of nuclear history is
traced in Great Britain and compared to
worldwide advancements in the nuclear industry. Non-technical discussions are provided
of the Magnox and AGR British reactors, the
light-water ones, PWR and BWR of the U.S.,
and the CANDU heavy-water reactor of Canada.
A brief background is given on how and why
these types of reactors were chosen by the
respective countries. The majority of the
book deals with the history of the
Advanced Gas-Cooled Reactor (AGR) and why it
was "best" for Great Britain. The latter
portion of this work covers the controversial
Windscale Decision dealing with the expansion
of a nuclear processing plant. Although
this has a purely British angle to it, *The
Nuclear Power Decisions* can serve as a
valuable historical guide to global decision
making on nuclear concerns.

31.   York, Herbert F.   THE ADVISORS: OPPENHEIMER,
         TELLER, AND THE SUPERBOMB.   New York:
         Freeman, 1976.

              With his insight and wisdom as father of
         the nuclear bomb, J. Robert Oppenheimer
         recommended in 1949 that the Atomic Energy
         Commission not develop a hydrogen bomb. His
         arguments were social, political, and techno-
         logical, but he lost to a bright physicist,
         Edward Teller. The result was the develop-
         ment of the hydrogen bomb by 1952 followed
         three years later by the same success in the
         Soviet Union. This is a superb memoir of
         two men who were paramount in determining
         the nuclear course of America. Any guide
         to the nuclear debate should include a
         history of the Teller-Oppenheimer connection
         even though this one is somewhat technical.

Anti-Nuclear

32. Aldridge, Robert C.   FIRST STRIKE!: THE
    PENTAGON'S STRATEGY FOR NUCLEAR WAR.
    Boston: South End Press, 1983.

    In the last few years the number of
    scientists who once earned their living in
    the nuclear industry but who have now turned
    against nuclear power has increased drama-
    tically. Aldridge is one of these indivi-
    duals and up until 1973 was an engineer for
    Lockheed Missiles & Space Co. *First Strike!*
    is an insider's critique of the military-
    industrial complex which adheres to the
    philosophy that the U.S. is leaning toward
    a first strike capability. The book goes
    into great detail about ballistic missiles,
    MX and cruise missiles, and other forms of
    nuclear-based military technology. Written
    for the general interest reader, Aldridge's
    perspective lends much to the understanding
    of why so many nuclear industry workers have
    turned sour on nuclear power.

33.   ATOM'S EVE: ENDING THE NUCLEAR AGE, AN
       ANTHOLOGY. Edited by Mark Reader,
       Ronald Hardert, and Gerald L. Mouton.
       New York: McGraw-Hill, 1980.

           More than forty, fiery contributed essays
       make for lively reading in this anti-nuclear
       book. Some of the comments are very brief
       and others quite lengthy, but all have one
       premise--Earth will be uninhabitable or
       destroyed if something is not done now
       about nuclear power and its waste. Every-
       thing from Three Mile Island to health
       problems to nuclear war to cover-up cases is
       discussed. Some of the more notable contri-
       butors include Helen Caldicott, Barry
       Commoner, Jacques Cousteau, Vernon Jordan,
       Amory B. Lovins, and Lewis Mumford. An
       extensive checklist of energy alternatives
       and organizations, a calendar of major
       nuclear events since 1945, and a bibliography
       of nuclear books and films are appended.

34.   Beres, Louis Rene. TERRORISM AND GLOBAL
       SECURITY; THE NUCLEAR THREAT. Boulder,
       CO: Westview, 1979. (Westview Special
       Studies in National and International
       Terrorism)

           This is a political scientist's view
       of modern terrorism which points out that
       uncontrolled terrorism could pose a nuclear
       threat. Beres talks about preventive
       security measures, how to deal effectively
       with terrorist behavior, and he provides
       readers with his own preventive plan of
       international cooperation. The book is
       entirely devoted to the threat of terrorism
       and does not include discussions of the
       technical aspects of nuclear power. On the
       socio-political level this is an eloquently
       written and thought-provoking work directed
       toward the well-read nuclear individual.

35.   British Medical Association.   Board of
       Science and Education.   THE MEDICAL
       EFFECTS OF NUCLEAR WAR: THE REPORT OF
       THE BRITISH MEDICAL ASSOCIATION'S BOARD
       OF SCIENCE AND EDUCATION. New York:
       Wiley, 1983.

       Ever since the effects of nuclear war
       have been studied, many people have argued
       that all that there is to know about the
       topic is already known. However, as this
       book points out, the 1960s and 1970s saw
       nuclear testing malfunctions and the ozone
       layer controversy, both undetected after-
       effects. Only a year ago the situation of
       reduced sunlight due to the amount of soot
       in the atmosphere was discussed and may
       turn out to be the most detrimental of all
       nuclear war effects. The Board questions
       the governmental estimates of blast
       casualties and offers its own predictions.
       The example used is that of a Soviet attack
       on the UK of about 200 megatons, or over
       three tons of explosives for every inhab-
       itant of the British Isles. The level of
       the attack could be higher if cruise
       missiles are used. The number of fatali-
       ties resulting from such an assault is
       estimated as between twenty-five and forty
       million. The international medical pro-
       fession is concerned about these conse-
       quences mainly because physicians are now
       being asked to assist in the planning for
       the aftermath of a nuclear war. Despite
       the conflicting predictions of the official
       government sources and the private scienti-
       fic studies, the conclusion of the group
       remains the same--the nation's health
       officials and practitioners could not
       handle the casualties from just a one
       megaton attack.

36.   Berger, John J. NUCLEAR POWER--THE
        UNVIABLE OPTION: A CRITICAL LOOK AT OUR
        ENERGY ALTERNATIVES. Introduction by
        Linus Pauling. Palo Alto, CA: Ramparts,
        1976.

        Beginning with the ABCs of reactor
        technology, this book progresses through
        discussions of the spiraling costs of
        nuclear power, the radioactive garbage it
        leaves behind, and the generally known
        risks involved. The impetus of the work
        is, however, to provide viable alternatives
        to nuclear power. These other sources of
        power include solar power, wind, geothermal,
        ocean and tidal power as well as coal
        gasification. The solution offered is that
        only a citizen's movement can halt the
        government supported nuclear growth in
        favor of the many alternatives.

37.   Briggs, Raymond. WHEN THE WIND BLOWS.
        New York: Schocken Books, 1982.

        In a whimsical, cartoon format this
        scenario reveals an elderly English
        couple who have just heard of an impending
        nuclear attack. Like trained poodles they
        follow the vague governmental guidelines
        for survival and only live through the
        first 48 hours of the attack. The inner-
        most thoughts, musings, and questions
        raised by this simple, peasant couple
        provide much food for thought. This is a
        most gruesome, humorous book with an
        adult message.

38.   Burleson, Clyde W. THE DAY THE BOMB FELL
       ON AMERICA: TRUE STORIES OF THE
       NUCLEAR AGE. Englewood Cliffs, NJ:
       Prentice-Hall, 1978.

       This collection of documented stories
       recounts some bizarre accidents or near-
       accidents involving nuclear or atomic
       weaponry worldwide. Most of the stories
       first appeared in the newspapers and were
       quickly discredited by U.S. and foreign
       governments as not being critical or in
       some cases, not ever occurring. Using
       these incidents as a backdrop the author
       demonstrates why we have the right to dis-
       trust nuclear power. Included are details
       of Soviet atomic disasters, the Karen
       Silkwood incident, and the invasion of a
       U.S. citizen's rights by a nuclear energy
       organization. The author offers a sane
       approach for coping with atomic energy, a
       power source which he realizes is probably
       here to stay.

39.   Caldicott, Helen. NUCLEAR MADNESS: WHAT
       YOU CAN DO! Brookline, MA: Autumn
       Press, 1978.

       This pediatrician turned nuclear acti-
       vist believes that life on our planet will
       become extinct if the progress of nuclear
       technology is not halted. Everything
       around us will become contaminated thus
       threatening the lives of everyone if radio-
       active pollutants remain unchecked. Dr.
       Caldicott makes an emotional plea for all
       individuals to become involved in the
       campaign to eliminate nuclear power. The
       U.S. cannot trust its elected officials to
       do the job because most of them have no
       sense of moral responsibility for humanity's
       future and their desire for re-election
       influences their decisions. Each citizen
       needs to be his own committee of one and
       help to avert M.A.D.ness--Mutually
       Assured Destruction.

40.   Clayton, Bruce D. LIFE AFTER DOOMSDAY.
        Boulder, CO: Paladin, 1980.

        This is a survivalist book which
        provides step-by-step plans for living
        through a nuclear war. Information about
        shelters, food storage, home medical
        techniques, and the psychology of survival
        are given. The author informs you of the
        prime target areas of the U.S. as well as
        those least likely to be hit. This is a
        most comprehensive work on survival which
        would be useful to follow through any
        major natural or man-made disaster.

41.   Committee for the Compilation of Materials
            on Damage Caused By the Atomic Bombs
            in Hiroshima and Nagasaki. HIROSHIMA
            AND NAGASAKI: THE PHYSICAL, MEDICAL
            AND SOCIAL EFFECTS OF THE ATOMIC
            BOMBINGS. Trans. by Eisei Ishikawa
            and David Swain. New York: Basic,
            1981.

        As a large, photographic essay book
        Hiroshima and Nagasaki makes a powerful
        statement against the power and destruction
        potential of a nuclear attack. The Japan-
        ese have prepared this book as a historical
        account of everything related to atomic
        bombs and as a graphic presentation of the
        first, and hopefully the last, bombs of
        this type. Medical documents report that
        the first case of leukemia appeared as
        early as 1945 and that an atom bomb induced
        cataract was treated in 1948. Various
        cancers of the thyroid, breasts, and lungs
        also increased after the war. As yet no
        harmful genetic effects have been observed
        but an on-going generation by generation
        study is still being conducted. The topic
        of psychological stress is touched upon
        and as one scarred and burned victim puts
        it "Unless you are an A-bomb victim you
        can't understand." The survivors of the

attack suffer greatly and they feel that
the Japanese government refused to assist
them and even to this day, ignores them.
It has rejected the demand for an A-bomb
Victims Relief Law that would provide aid
to those who suffered bodily injury or loss
of livelihood. To the residents of these
two cities the Emperor's surrender within
ten days of the bombings was more shocking
than the attack itself as this made them
feel completely abandoned and rejected.
This massive volume provides a sombre look
at the living dead of Hiroshima and Nagasaki.

42.   COUNTDOWN TO A NUCLEAR MORATORIUM.
            Richard Munson, ed. Washington, DC:
            Environmental Action Foundation, 1976.

       Nuclear power is far too costly in
terms of the money spent, the environment
wasted, and the safety hazards which exist.
This is the primary thesis of this book of
fourteen essays written by well-known
anti-nuclear scholars, environmentalists,
doctors, and politicians. Coverage
includes the Browns Ferry incident, nuclear
safety, alternatives to nuclear power, and
the case for a nuclear moratorium. In
1979, after Three Mile Island, an expanded
version of this book was published under
the title *Accidents Will Happen: The Case
Against Nuclear Power*. (Item number 49)

43.   Cox, Arthur M. RUSSIAN ROULETTE: THE
        SUPERPOWER GAME. New York: Times
        Books, 1982.

        With present day sophisticated computer
systems set to launch nuclear missiles on
warning, our survival can no longer be
assured. We have previously survived false
alarms--the risk now is that we could be
annihilated by accidental nuclear war. Cox
places the blame on the hawks in Moscow and
Washington. This work shows how the
policies of the two countries reinforce
that of each other. To avoid annihilation,
negotiation is necessary. Cox presents a
proposal for a two-track negotiated solution.
One solution bans direct or indirect military intervention by Third World superpowers
and the second is based on an agreement to
end the nuclear arms race. Cox draws upon
his extensive background in diplomacy,
intelligence, and arms control to suggest
how we may be able to divert this disastrous
course. A Soviet commentary is provided by
Georgy Arbatov, the foremost Soviet expert
on the U.S. and a close advisor to
President Reagan.

44.   Croall, Stephen. THE ANTI-NUCLEAR HANDBOOK.
        New York: Pantheon, 1978.

        Here is a most unusual comic book style
of documentary which warns against construc-
ting nuclear power plants, nuclear waste
dumps, of possible sabotage, and the
potential for accidents. Some historical
figures are brought to life in vignettes in
which their original words have been re-
written to reflect nuclear energy messages.
Most all of the points of the nuclear
controversy are included in this book but
their presence is masked by the cartoons.

45.   Curtis, Richard, and Elizabeth Hogan.
       NUCLEAR LESSONS: AN EXAMINATION OF
       NUCLEAR POWER'S SAFETY, ECONOMIC, AND
       POLITICAL RECORD. Harrisburg, PA:
       Stackpole, 1979.

            Portions of this book appeared under
       the title *The Perils of the Peaceful Atom:
       The Myth of Safe Nuclear Power Plants.*
       published in 1969. This work points out
       the fact that some of the authors' pre-
       dictions came true with Three Mile Island.
       The authors provide a thirty-year history
       of nuclear lessons and reiterate their
       beliefs that there are more practical and
       safe alternatives to the energy question
       than nuclear power.

46.   _____. THE PERILS OF THE PEACEFUL ATOM:
       THE MYTH OF SAFE NUCLEAR POWER PLANTS.
       New York: Ballantine, 1969.

            The authors attempt to bring together
       in one source the facts and figures which
       will alert U.S. citizens to the risks,
       problems, and errors of nuclear power.
       In their plea for a re-examination of our
       peaceful nuclear program they argue that
       the safety of every industry is now hanging
       on that of the nuclear industry and warn
       that commercial atomic power is an
       unpromising investment. Utilizing the
       scare tactic technique, the authors cover
       the full gamut of the nuclear question,
       cite hundreds of cases and experts, and
       provide readers with their predictions
       of the future of nuclear generated power.

47.   Drinian, Robert F.   BEYOND THE NUCLEAR
        FREEZE.   New York: Seabury, 1983.

        In this brief, simple book Father
        Drinian touches upon the topics of weapons
        proliferation and government policy. As a
        former congressman and law school professor
        he lends a welcomed insight to the politics
        of nuclear power and as a religious leader
        he displays his moral concerns. The book
        follows the growing anti-nuclear movement,
        the efforts to negotiate a nuclear mora-
        torium, and discusses freeze strategies.
        Father Drinian concludes that due to its
        global position and foreign policy, the U.S.
        will remain a nuclear power, but that
        activist citizen's groups can still strive
        to remove the threat of nuclear annihila-
        tion.

48.   Dunn, Lewis A.   CONTROLLING THE BOMB:
        NUCLEAR PROLIFERATION IN THE 1980s.
        New Haven, CT: Yale University Press,
        1982.

        Dunn desires to slow the spread of
        nuclear know-how and weapons but does not
        believe that further proliferation will
        have beneficial effects. Pointing a finger
        at the strife in the Middle East, he
        wonders what the situation would be (or
        have been) if Israel, Syria, Pakistan and
        others possessed nuclear warfare. Either
        by theft, sabotage, outright purchase or by
        their own technological expertise, Dunn
        predicts that other countries will be able
        to enter the now elite nuclear club. He
        attacks the misconception that the presence
        of nuclear weapons may bring peace and
        stability to the Middle East by establishing
        regional deterrants much like that which
        prevented a US-USSR war. The latter half
        of *Controlling the Bomb* discusses a range
        of alternatives for a U.S. anti-prolifera-
        tion strategy. Dunn believes that the U.S.

should re-establish itself as a reliable
supplier of nuclear fuel, take aim to close
the technology export loopholes, impose
military and diplomatic sanctions on proliferation firebreaks as he calls them,
behind which nations could be held by the
threat of sanctions. This is an in-depth
political and diplomatic study of the
nuclear question which will be appreciated
by the knowledgeable layperson.

49.   Environmental Action Foundation. ACCIDENTS
        WILL HAPPEN: THE CASE AGAINST NUCLEAR
        POWER. Introduction by Ralph Nader.
        New York: Perennial/Harper, 1979.

        Here is a collection of essays from a
        cross-section of America including senators,
        doctors, lawyers, and enraged citizens who
        lash out against the nuclear industry.
        Topics covered include a history of radio-
        active wastes, long-term effects of radia-
        tion, and discussions of what happened at
        Browns Ferry and Three Mile Island. These
        are sincere and well-written essays. This
        book is an expanded version of the 1976
        title *Countdown to a Nuclear Moratorium*.
        (Item number 42)

50. Faulkner, Peter, ed. THE SILENT BOMB: A
     GUIDE TO THE NUCLEAR ENERGY CONTROVERSY.
     New York: Random, 1977.

     Nuclear power represents the greatest
     single threat to the health and safety of
     humanity. This is the overriding premise
     of this powerful guide to the atomic
     industrial complex. Numerous accusations
     of U.S. government and business involvement
     in uranium price fixing, the placing of
     pro-nuclear people in key government posi-
     tions, and the concealment of studies
     whose results showed the nuclear risks are
     spelled out. The major portion of the book
     is comprised of excerpts from previously
     published books and articles written by
     well-known people, some of whom are
     represented elsewhere in this bibliography.

51. THE FINAL EPIDEMIC: PHYSICIANS AND
     SCIENTISTS ON NUCLEAR WAR. Ruth Adams,
     and Susan Cullen, eds. Chicago: Educa-
     tional Foundation for Nuclear Science,
     1981.

     Twenty-two writers contribute essays to
     this book which recognizes the emergence of
     a new militarism in American thought.
     Exaggerated fears of the Soviet
     Union in recent years are to blame for our
     reckless foreign policy and provide the
     impetus for these comments by leading
     scientists and doctors. Coverage includes
     assessments of burn casualties, diseases,
     radioactive fallout, and cancer in atomic
     bomb survivors. The book closes with
     suggestions for preventing nuclear war and
     a statement by the International Physicians
     for the Prevention of Nuclear War.

52.   Ford, Daniel. BEYOND THE FREEZE: THE ROAD
       TO NUCLEAR SANITY. Boston: Beacon, 1982.

       The issue of a nuclear freeze is a simple,
uncomplicated notion: "the United States
and the Soviet Union should agree to a dead
halt in all aspects of the nuclear arms
race." This also entails the cessation of
further nuclear testing of any nature. This
proposal is discussed in great detail and
attempts are made to answer many pressing
questions. Excellent analyses are made as
to the key developments in nuclear arms,
the build-up of nuclear weaponry by the U.S.,
and a review of the pros and cons of the
freeze proposal.

53.   _____. THE CULT OF THE ATOM: THE SECRET
       PAPERS OF THE ATOMIC ENERGY COMMISSION.
       New York: Simon and Schuster, 1982.

       Mr. Ford is a former executive director
of the Union of Concerned Scientists, a
well-known anti-nuclear lobby group. He
contends that the Atomic Energy Commission
is guilty of a cover-up with regard to
nuclear safety in the U.S. To prove his
point the author took advantage of the
Freedom of Information Act and read hundreds
of AEC "secret" reports. The Emergency Core
Cooling System (ECCS) hearings of 1971-72
and the Reactor Safety Study of 1975 are
extensively covered. Ford recommends up-
grading the safety of existing plants and
even suggests that the operation of some
located in heavily populated areas be cur-
tailed. His arguments add powerful fuel to
the popular and political anti-nuclear
forces. In general, this work attacks
federal programs and agencies, the practice
of poor site selection, and of course, the
individuals surrounding the controversy.

54.   _____. THREE-MILE ISLAND: THIRTY MINUTES
           TO MELTDOWN. New York: Viking, 1982.

           An overview of the TMI accident and an
           explanation of why this is considered to
           have been a "Class 9" accident are provided
           for readers. The author sees some value
           in the incident in that it provided a good
           test case for all to study as well as
           having been an exercise in crisis manage-
           ment. Also covered are the NRC's plans for
           self-improvement and an analysis of its
           policies regarding nuclear safety. Some
           serious problems still exist with NRC
           reports and paperwork. Answers to
           questions raised by a presidential commis-
           sion established to look into the TMI
           incident have been suppressed by the Carter
           and Reagan administrations. Speculations
           as to why this occurred are offered.

55.   Fuller, John Grant. WE ALMOST LOST DETROIT.
           New York: Reader's Digest, 1975.

           Beginning with tales about the accident
           at Fermi No. 1 liquid metal fast-breeder
           reactor (LMFBR) in Lagoona Beach, Michigan
           in 1966, the author dramatically unfolds
           the safety issues surrounding nuclear
           reactor technology. Numerous other acci-
           dents or near-accidents not generally
           known to the public are also explained.
           The primary thrust of the book is to reveal
           the dangers of nuclear power and the fact
           that the public was misled by the former
           Atomic Energy Commission into thinking that
           nuclear energy is safe. The author con-
           tends that the AEC was so frightened by
           the estimated consequences of an accident
           that the members did not want to reveal
           them to the public. This is a fascinating
           book whose findings and arguments can help
           the cause of the anti-nuclear movement.

56.   Gofman, John W.   "IRREVY": AN IRREVERENT,
       ILLUSTRATED VIEW OF NUCLEAR POWER: A
       COLLECTION OF TALKS, FROM BLUNDERLAND
       TO SEABROOK IV.   San Francisco:
       Committee for Nuclear Responsibility,
       1979.

       This is one of the more humorous and
       satirical books in the nuclear genre.
       Using cartoons to illustrate his points
       and issues, Dr. Gofman puts into print nine
       of his speeches given between 1975 and 1978
       on the virtues of life without nuclear
       power. Numerous facts and figures are
       provided to lend some credibility to the
       cartoons.

57.   _____, and Arthur R. Tamplin.   POISONED
       POWER: THE CASE AGAINST NUCLEAR POWER
       BEFORE AND AFTER THREE MILE ISLAND.
       Emmaus, PA: Rodale, 1979.

       In 1971, eight years before the Three
       Mile Island incident, these two authors
       predicted this near-disaster in a book
       entitled *Nuclear Power*. This revised and
       much expanded edition of that book revolves
       around the TMI affair. Great concern is
       still expressed for the health of future
       generations, the cover-ups which are
       commonplace in the government and the
       nuclear industry, and the controversial,
       utility-protecting Price-Anderson Act of
       1957. In arguing that nuclear power is not
       necessary to produce electricity since
       today nuclear power only supplies 3½% of
       our total energy needs, the authors speak
       in favor of the alternative sources of
       energy. The book has many useful charts,
       diagrams, and statistics to support its
       arguments against nuclear power.

58.   Goodwin, Peter. NUCLEAR WAR: THE FACTS ON
         OUR SURVIVAL. New York: Rutledge,
         1981.

         Nuclear War provides readers with an
         insight into how the quest for atomic
         energy has resulted in a massive stockpile
         of nuclear bombs worldwide. Goodwin
         clearly explains nuclear power and shows
         who has it and who is most likely to use
         it in case of war. Coverage is provided
         which shows how nuclear radiation affects
         our environment and people. A unique chap-
         ter in this book is devoted to the pros
         and cons of evacuation, home modification,
         building and shelter types, and what
         supplies should be maintained in storage
         for survival purposes. This easy-to-read,
         well-illustrated book is a plea for reason
         and understanding that anything which
         lessens the danger of war is good.

59.   GRASSROOTS: AN ANTI-NUKE SOURCE BOOK.
         Fred Wilcox, ed. Trumansburg, NY:
         Crossing, 1980.

         Fifty-six very brief articles comprise
         this guide to protesting against nuclear
         power in very radical ways. Included is a
         checklist for demonstrators, a guide on
         how to use civil disobedience to obstruct
         nuclear power, tips on how to defend your-
         self without a lawyer, how to petition,
         and a variety of other ingenious plans.
         This "how to" book also has extensive lists
         of people, organizations, and newsletters
         to which you can turn for support and
         information. For financial assistance a
         list of foundations which grant money to
         combat the spread of nuclear power is most
         helpful.

60.   Gyorgy, Anna and Friends. NO NUKES:
      EVERYONE'S GUIDE TO NUCLEAR POWER.
      Boston: South End, 1979.

      Numerous experts worldwide have con-
      tributed to make *No Nukes* one of the better
      books of the anti-nuclear genre. It really
      is a guidebook leading the reader from a
      history of atomic development to the
      complex economics to the alternative
      sources of energy with which we should
      learn to live. Two revealing chapters
      examine the nuclear power structure and
      the existence of planned development of
      nuclear power in foreign countries. The
      book has many useful statistics, thorough
      and extensive references, and exceptional
      photos and drawings. This is another
      good "first book" on nuclear power.

61.   International Physicians for the Prevention
      of Nuclear War. LAST AID: THE MEDICAL
      DIMENSIONS OF NUCLEAR WAR. (Papers
      from a Congress) Eric Chivian, ed.
      San Francisco: Freeman, 1982.

      Seventy-two physicians from twelve
      countries attended the First Congress of
      the International Physicians for the Pre-
      vention of Nuclear War during which time
      they discussed how they may help to
      decrease the costs of the arms race (eco-
      nomic, moral, and psychological) by speak-
      ing out against nuclear war. The president
      of the group, Barnard Lown, states that
      world arms spending is around $1.4 billion
      per day, or $1 million per minute. He
      argues that a diversion of just three weeks
      of these funds, or about $30 billion, could
      provide a sanitary water supply for the
      entire world. Not only physicians, but all
      people should be concerned with the
      enormous spending for nuclear arms and try
      to persuade world leadership to spend money

to save lives, not to destroy them. In the
summary readers are reminded that "the
genocidal nature of nuclear weapons has
rendered nuclear war obsolete as a viable
means for resolving conflict....Wars begin
in the mind, but the mind is also capable
of preventing war."

62.   Jungk, Robert. THE NEW TYRANNY: HOW
       NUCLEAR POWER ENSLAVES US. New York:
       Grosset & Dunlap, 1979.

       Originally published in Germany in 1977,
this book has been credited with having a
major role in the banning of nuclear energy
in Austria and possibly Switzerland. It
has also fueled the fire of the nuclear
controversy in France, Germany, and Sweden.
Jungk logically explains that by following
the path of nuclear power we will be forced
to surrender some liberties one at a time
until we are finally part of a regimented
society. The book is complete with a
record of nuclear accidents and failures,
discussions of the proliferation and
terrorism questions, and the role of politics in this global concern for nuclear
power.

63.   Kaplan, Fred. THE WIZARDS OF ARMAGEDDON.
       New York: Simon & Schuster, 1983.

       Many writers, activists, and research-
ers have taken advantage of the recently
declassified government documents to
investigate the nuclear question. Journal-
ist Kaplan did just this to prepare a
historical development of nuclear weapons
policy since World War II. The book also
serves as a mini-biography of the chief
nuclear scientists of the time period and
reveals life as it was inside the labora-

tories which were cloaked in secrecy for
nearly three decades. Briefly considered
is the Reagan administration's policy and
views toward nuclear power and nuclear war.
Kaplan believes that President Reagan
supports limited nuclear war fighting but
adds that nobody in the administration has
ever asked how a nuclear war might be
fought. The author thinks that the question
should be asked and that the various
alternatives should be studied.

64. Katz, Arthur M. LIFE AFTER NUCLEAR WAR:
    THE ECONOMIC AND SOCIAL IMPACTS OF
    NUCLEAR ATTACKS ON THE UNITED STATES.
    Cambridge, MA: Ballinger, 1982.

    This lengthy book takes issue with the
    terms "acceptable" and "unacceptable"
    damage as they pertain to measures of
    psychological destruction and survival
    after a nuclear weapons exchange. Katz
    informs readers that the amount of weapons
    required to create unacceptable damage is
    far smaller than we have been led to
    believe. Were key industrial facilities
    to be rendered inoperable by well-planned
    attacks, a modern industrial society would
    be seriously crippled. We are told that
    the social, political, and economic life
    would be so drastically altered in an
    attack that most of us would consider it to
    be unacceptable damage. Katz adds that all
    of this doesn't even take into consideration
    the possibility that even minimal key hits
    could prevent any type of biological sur-
    vival. Katz also criticizes the high level
    crisis evacuation plans as useless. In-
    stead of developing these plans, high
    government officials should be concerned
    with developing plans for a nuclear disar-
    mament and a nuclear freeze.

65.   Langer, Victor, and Walter Thomas.
       THE NUCLEAR WAR FUN BOOK. New York:
       Holt, Rinehart and Winston, 1982.

            In a morbid, yet humorous manner, the
       authors present an illustrated book of fun
       and games to be enjoyed after a nuclear
       holocaust. As they state, "Just because
       you may be surrounded by destruction,
       doesn't mean you can't be creative."
       Included are post-nuclear war activities
       and a few pre-war games. Some can be
       played outdoors after the radiation level
       has dropped (Radioactive Tag) and others
       must be played in the fallout shelter
       (Connect the Craters and Backgamma Ray).
       This is definitely not a book you would read
       to your children at bedtime.

66.   Leppzer, Robert, ed. VOICES FROM THREE
       MILE ISLAND: THE PEOPLE SPEAK OUT.
       Trumansburg, NY: Crossing, 1980.

            Full of emotion, this small book
       includes testimonials from thirteen
       residents in the TMI accident area. In
       a folksy style we hear what each was doing
       when the accident occurred, their immediate
       thoughts, and how their lives have been
       affected since that March 1979 day. The
       worries, frustrations, and the anger of
       each resident is felt as you read these
       interviews. Most realize that there is a
       need for energy but now they think there
       may be safer ways to generate it than
       through nuclear power.

67.   Lovins, Amory B., and L. Hunter Lovins.
       ENERGY/WAR: BREAKING THE NUCLEAR LINK.
       San Francisco: Friends of the Earth,
       1980.

          Nuclear power equals nuclear war as we
       are told in some very convincing arguments
       by the Lovinses. The U.S., and others with
       the atomic knowledge, had best not share
       this knowledge by educating foreign
       students in our universities. We are
       reminded that nuclear power is not a sig-
       nificant substitute for oil. Nuclear
       power only generates electricity which
       amounts to one-tenth of the U.S. power
       needs. Oil, on the other hand, has numer-
       ous household and industrial uses. The
       authors provide their plans for saving oil
       and offer some advice for Detroit auto-
       makers. This is not an elementary book on
       the nuclear power issue but, nevertheless,
       the writings of this well-known husband and
       wife anti-nuclear team will be appreciated
       by the well-informed reader.

68.   Nader, Ralph, and John Abbotts. THE MENACE
       OF ATOMIC ENERGY. New York: Norton,
       1977.

          In a manner of writing that only he can
       get away with, Ralph Nader takes on the
       nuclear power industry, its scientists,
       the governmental agencies, and the
       utilities which support and manufacture
       reactor technology. Pointing to the fact
       that nuclear engineers have resigned due to
       the defects and deficiencies in the nuclear
       industry and the regulatory agencies, Nader
       surmises that the reactor industry is
       crumbling technically and economically.
       Mr. Nader vehemently exclaims that atomic
       fission is unsafe, unnecessary, and an
       economic folly in this controversial book.

69.   Neild, Robert. HOW TO MAKE UP YOUR MIND
        ABOUT THE BOMB. London: Deutsch, 1981.

        This Cambridge University Professor of
        Economics uses this small book to
        contribute to the British policy debate
        regarding American nuclear presence in
        Great Britain. He contends that Britons
        are more threatened than strengthened by
        American nuclear bases and further adds
        that Britain may not need them. The United
        States can be relied upon to defend its
        investments and role in Europe but the
        presence of nuclear weapons on British soil
        adds to the risk that Britain will be
        subject to a nuclear attack. Neild notes
        that U.S. assistance to the British nuclear
        weapons program has had two conditions:
        that British forces be assigned to NATO and
        that Britain continues to provide the U.S.
        with military facilities. Fervent in his
        displeasure with this agreement he calls
        for a public debate about U.S. military
        bases in Britain, asks where the nuclear
        arms race is going, and sees dangers in
        superpower arms control negotiations.
        Neild wants Britain to become self-reliant
        in matters of defense like the French, but
        without nuclear weapons. With this book
        the author makes frequent attacks on the
        reckless American foreign policy.
        Although this is geared to British reader-
        ship, Americans will find it useful as a
        history of U.S. nuclear involvement over-
        seas and as a guide to understanding the
        strategic importance of European military
        bases.

70.   Nelkin, Dorothy, and Michael Pollak.
        THE ATOM BESIEGED: ANTI-NUCLEAR
        MOVEMENTS IN FRANCE & GERMANY.
        Cambridge, MA: MIT Press, 1982.

        Focusing on the problems in western
        Europe, Nelkin and Pollak provide a

comparative analysis of nuclear opposition
in France and Germany. Recent explosive
and emotional attacks have been made in
both countries by the anti-nuclear movement.
A history of the movement and the valuable
governmental responses provides us with an
understanding of the nuclear opposition and
a look at the political processes of two
European countries. This is not a book
for the novice nuclear reader but rather
it is directed to the well-informed
individual.

71.   NUCLEAR WAR: WHAT'S IN IT FOR YOU?
        New York: Pocket Books, 1982.

        Ground Zero is the Washington, DC
based group responsible for putting
together this book of nuclear war advice.
The group is a nonpartisan educational
organization dedicated to supplying information which the public has a right to
know. The book covers the immediate
effects of nuclear war and what to do if
you do manage to survive a nuclear
exchange. A scouting report on the
Russians, a game of twenty questions about
nuclear war, and the weapons proliferation
problem round out the book.

72.   O'Keefe, Bernard J. THE NUCLEAR
       HOSTAGES. Boston: Houghton Mifflin,
       1983.

            Having worked with Robert Oppenheimer
       on the first atomic bomb, Mr. O'Keefe
       lends his privileged insight into global
       discussions on the nuclear advances since
       World War II to the general public. He
       provides an excellent history of fission
       research and of the Los Alamos Laborator-
       ies. This author is one of the growing
       number of scientists who formerly made
       their living in the nuclear industry and
       who have now turned against it. Among his
       warnings he attacks the treaty process as
       being ineffective in stopping the arms
       race. O'Keefe has his own survival plan
       which he is willing to share--"Stay home,
       take two aspirins, and hope for the best."
       The hostages referred to in the title are
       the Russian and American citizens who are
       held hostage to each other. According to
       the author we are held hostage to the
       mutual strategies of deterrence or
       mutually assured destruction.

73.   Olson, McKinley C. UNACCEPTABLE RISK: THE
       NUCLEAR POWER CONTROVERSY. New York:
       Bantam, 1976.

            The "technological monster", nuclear
       fission power, is once again on trial in
       this book which warns that nuclear power
       is a threat to the future of life on this
       planet. Four major concerns prevail
       throughout the text: 1) radioactive waste;
       2) the unexplored relationship between
       radioactivity, disease, and genetic defects;
       3) the dangers of error, sabotage, acci-
       dents, and theft; and 4) the question of
       a worldwide proliferation of nuclear
       weapons. A careful analysis is given to
       numerous nuclear plants and to the contro-
       versial roots of nuclear power. Many

examples of what citizens have done to
prevent nuclear power from entering their
neighborhoods and some predictions on what
the future will bring are also provided.
A nationwide list of nuclear plants, their
capacity, ownership, and operation date
is most useful.

74.   Peterson, Jeannie, and Don Hinrichsen, eds.
        NUCLEAR WAR: THE AFTERMATH. New York:
        Pergamon, 1982.

        This collection of fourteen articles
first appeared as a special issue of the
international journal, *Ambio*. The book
was published in the belief that knowledge
of the consequences of a nuclear war may
serve as a deterrent to such an event.
This is a very scientific, easy-to-read
appraisal of the effects of a global war
based not on the worst possible war, nor
even a "limited war," but one based on
the premise that once a nuclear war
breaks out it would probably be neither
containable nor controllable.

75.   Pollard, Robert, ed. THE NUGGET FILE.
           Cambridge, MA: Union of Concerned
           Scientists, 1979.

               The Freedom of Information Act
           allowed the Union of Concerned Scientists
           access to public Nuclear Regulatory
           Commission documents, one of which was
           "The Nugget File." This file was a
           special, internal file maintained by
           Dr. Stephen H. Hanauer, a senior NRC
           official, for over ten years. The file
           identifies serious accidents and safety
           deficiencies at U.S. nuclear power plants
           in very revealing, but brief, reports.
           Included are such deficiency reports as
           "Emergency Core Cooling System Inopera-
           tive-Valves in Wrong Position", "Loss of
           Control of Reactor Water Level-Blown Fuse",
           "Reactor Water Flow Obstruction-Plywood
           Left in Emergency Cooling Pipe" and many
           more. The Nugget File is too large to be
           published in its entirety so this book of
           abstracts provides researchers with an
           adequate summary and appropriate citations
           to the actual source should a more in-depth
           reading be necessary.

76.   Pringle, Peter, and James Spigelman.
           THE NUCLEAR BARONS. New York: Holt,
           Rinehart and Winston, 1981.

               In this lengthy book the secrets of
           the nuclear barons, the scientific elite
           who controlled the growth of the nuclear
           industry for over four decades, are
           revealed. This international study in
           radiation hazards, waste disposal, plant
           safety, and weapons proliferation provides
           much for the reader to ponder. The role
           of the British, French, and Indian govern-
           ments are chronicled as they pertain to

the development of nuclear power. The
authors do not advocate the shutdown of
all nuclear power plants in all nations
for they do realize that some nations may
have no practical alternative due to the
availability of resources. They do agree
that although nuclear power may be
necessary, its use should be kept to a
minimum. It is hoped that the information
in this book will help to bring about a
much needed change in the industry.

77.   PROTEST AND SURVIVE. E.P. Thompson, and
        Dan Smith, eds. New York: Monthly
        Review, 1981. Introduction by Daniel
        Ellsberg.

        As the American version of a British
book with the same title, this work has
ten contributed essays. It provides a
purely British and European view of
American nuclear policy and international
diplomacy. Europeans fear that the U.S.
and Russia are going to "fight it out"
and that Europe will be the stage for the
battle. The dangers of a nuclear war and
an unlimited arms race must be discussed
openly. The lengthy Ellsberg introduction
"Call to Mutiny" is by itself worth
reading and *if* what he says is true, then
we should prepare for World War III.

78.   Rapoport, Roger. THE GREAT AMERICAN
        BOMB MACHINE. New York: Dutton, 1971.

            In a tongue-in-cheek manner, Mr.
        Rapoport takes exception to the entire
        nuclear defense system of the U.S. He
        attacks the lack of common sense in the
        deployment of antimissiles, the irony of
        SALT and AEC weapons production programs,
        and such popular tourist attractions as
        the Sandia nuclear bomb museum run by the
        Department of Defense. The solution to
        a safe and guaranteed future includes the
        dismantling of the weapons program and
        the great American bomb machine should be
        declared a public health hazard. Rapoport
        argues that Dr. Glenn Seaborg, the AEC
        chairman at the time, should have resigned
        and the author abhors the fact that
        Seaborg was paid $100,000 in 1955 by the
        U.S. government for the fine job he did
        on the plutonium separative process used
        to make the Nagasaki bomb. Although some
        of the material is dated, we are still
        involved in SALT and weapons have once
        again become a political question.

79.   Scheer, Robert. WITH ENOUGH SHOVELS:
        REAGAN, BUSH AND NUCLEAR WAR. New
        York: Random House, 1982.

            This Los Angeles Times reporter has
        discovered a reversal in the longstanding
        American assumption that nuclear war means
        mutual suicide. Now the prevailing
        thought is that American leaders have
        chosen to fight and win in a nuclear war.
        The leaders also expect that once the war
        is won that the U.S. will return to
        normal. The leaders believe that we are
        living in a pre-war and not a post-war
        world. The Soviets are bent on world
        conquest and the only way in which the

U.S. can meet this challenge is with the
determination to shrink the Soviet empire
and to alter Soviet society. Scheer
discovered Reagan's secret plan in a report
called the National Security Decision
Document which commits President Reagan to
the idea that a global nuclear war can be
won. Scheer exposes the deadly course of
the U.S. in this book which was written by
means of exclusive interviews with President
Reagan and several of his key staff and
cabinet.

80.   Schell, Jonathan. THE FATE OF THE EARTH.
       New York: Knopf, 1982.

       The Fate of the Earth is a published
collection of articles which originally
appeared in *The New Yorker*. Schell
discusses his concept of the "nuclear
predicament" as people's indifference or
apathy to the threat inherent in the
nuclear arsenals. The world, on the whole,
just doesn't think about them very much.
Schell would like to change this attitude
and attempts to do so by providing an
account of what the nuclear holocaust
might be like. He is probably very
accurate in his belief that the human life
would be extinguished or life on Earth
would be impossible due to the loss of the
ozone layer. One of the author's scenarios
depicts what would happen if a one
megaton bomb was dropped above the Empire
State Building. The conclusion to this
dramatic work is a chapter called "The
Second Death", the theme of which is that
worse than being killed is that the whole
species might be eliminated. Schell also
contrasts the notions of death and extinction, the doctrine of deterrence, and
political solutions in his closing remarks.

81.   SHUT DOWN: NUCLEAR POWER ON TRIAL.
          Summertown, TN: The Book Pub. Co.,
          1979.

          Shut Down is the transcript of the
          court testimony of two noted nuclear
          activists as they spoke at the October 2,
          1978 Federal Court hearing for the case
          Jeannine Honiker vs. Joseph M. Hendrie
          et al in Nashville, Tennessee. Mrs.
          Honiker's nineteen year old daughter con-
          tracted leukemia--the alleged cause of
          which was radiation from a nearby nuclear
          power plant. With the aid of Catfish
          Alliance and Farm Legal, a case against
          the Nuclear Regulatory Commission was
          prepared. The objective was to intervene
          in the licensing process of and to
          ultimately shutdown what was then the
          world's largest nuclear plant located in
          Hartsville, TN. The only point to be
          determined in this hearing was whether
          or not the judge had the jurisdiction to
          hear the case. On January 12, 1979 the
          case was dismissed for lack of that
          jurisdiction. Testimony in the hearing
          and the commentary following the tran-
          script are both vehement protests and
          emotional pleas to abandon nuclear power
          for the sake of our children and the
          future of this world.

82.   Sternglass, Ernest. SECRET FALLOUT: LOW-
          LEVEL RADIATION FROM HIROSHIMA TO
          THREE-MILE ISLAND. Rev. ed. New
          York: McGraw-Hill, 1981.

          Secret Fallout is an expanded version
          of the 1972 book by Dr. Sternglass
          entitled *Low-Level Radiation*. As in the
          earlier edition the author is still
          concerned with the health effects of low-
          level radiation and why the military, the
          nuclear industry, and some scientists

have tried to bury the facts. Using
examples from incidents at U.S. nuclear
power generating plants, the author
describes the effects that nuclear testing
and power plant radiation releases have
had, or will have, on the lives of the
people.

83.   Sweet, Colin, ed. THE FAST BREEDER
        REACTOR: NEED? COST? RISK?
        London: Macmillan, 1980.

        As the title implies, this work should
should analyze the pressing questions
of need, cost, and risk of fast breeder
reactors. In actuality it is a collection
of fourteen papers attacking FBRs and
nuclear power in general. Breeders were
originally designed to be a less expensive
manner of producing nuclear energy. The
only way to find out whether or not this
production is actually cheaper would be
to build large fast breeder reactors.
This work is very much against this type
of thinking and argues that radiation
hazards to the people may limit the
growth of nuclear power and that nuclear
energy and civil liberties are incompat-
ible. Other essays attempt to sway
readers to investigate more efficient
alternative sources like oil burning
heaters. Great concern is expressed for
the dangerous disposable wastes of the
fast breeder reactors. Finally, rather
than calling just for an end to FBRs,
this book calls for an end to all nuclear
power generation.

84.   Tamplin, Arthur R., and John W. Gofman.
          POPULATION CONTROL THROUGH NUCLEAR
          POLLUTION. Chicago: Nelson-Hall,
          1970.

          Two of the most knowledgeable anti-
          nuclear scientists attack the atomic
          industry with much fervor and emotion in
          this book. Citing criticism from within
          the nuclear industry and the fact that
          self-serving nuclear technology promoters
          are ignoring the known health hazards,
          both physicists call for a nuclear
          moratorium. Blind to all known good
          which has come from nuclear research, Drs.
          Gofman and Tamplin continue their crusade
          to halt the advancement of the industry.
          They freely use cynical phrases such as
          "Will your grandchild's genes be fit for
          your great-grandchildren to wear? " to
          make their points about radiation,
          pollution, and public health arrogance.
          This is a revealing easy-to-read book by
          two of the better known polemics in the
          campaign to thwart the growth of nuclear
          technology.

85.   Taylor, Theodore B., and Mason Willrich.
          NUCLEAR THEFT: RISKS AND SAFEGUARDS.
          Cambridge, MA: Ballinger, 1974.

          Starting off with a chapter on how
          to make nuclear weapons, the authors
          clearly make their point--anyone can
          obtain the formula through public, un-
          classified sources. Other sections of the
          book cover the U.S. safeguards against
          nuclear theft, the numerous types of
          nuclear theft risks and an abundance of
          recommendations which the Atomic Energy
          Commission, now the Nuclear Regulatory
          Commission, should consider. Useful
          tables and charts reflecting the growth
          of nuclear power complement the text.

86.   TO AVOID A CATASTROPHE: A STUDY IN FUTURE
        NUCLEAR WEAPONS POLICY. Michael P.
        Hamilton, ed. Grand Rapids, MI:
        Eerdmans, 1977.

        Twelve essays comprise this anthology
        of nuclear gloom and doom. A prevailing
        concern is that the proliferation of
        atomic weapons will continue and that a
        nuclear strike is inevitable. It is also
        hoped that such a strike will instill so
        much fear and concern in the people of
        this world that the aggressors, most
        likely the U.S. and/or Russia, will step
        back, look at what they have done, and
        take aim to ensure that it will not happen
        again. The topics of the papers include
        a history of the atomic age, a terrorist
        attack on a nuclear facility, sabotage,
        safety and security, and some practical
        steps toward disarmament.

87.   Touraine, Alain et al. ANTI-NUCLEAR
        PROTEST: THE OPPOSITION TO NUCLEAR
        ENERGY IN FRANCE. Trans. by Peter
        Fawcett. New York: Cambridge, 1983.

        The anti-nuclear movement in France is
        creating a new generation of educated and
        upperclass protestors and activists. This
        book is a study of several of those
        groups including industrial unionists,
        engineers, physicians, and housewives.
        Touraine and his social researchers
        arranged for several group discussions at
        which anti-nuclear individuals were
        gathered to speak with each other or to
        be challenged by others present who had
        opposing views. Discussions in the book
        between these diverse groups center
        around an idea exchange with the ultimate
        goal being the founding of a new social

movement within the anti-nuclear campaign.
This is exactly what Touraine and his
colleagues are trying to accomplish and
they try to direct these anti-nuclear
circles to more pragmatic forms of political action. The authors feel that these
groups are at the center of pending social
change and modernization. The thoughts of
many of those interviewed are fantasy-like,
making readers question the ideological
attitudes of these French citizens. This
is a superficial study that will provide
a minimal understanding of the French
resistance to nuclear growth.

88.   Union of Concerned Scientists. THE
        NUCLEAR FUEL CYCLE: A SURVEY OF THE
        PUBLIC HEALTH, ENVIRONMENTAL, AND
        NATIONAL SECURITY EFFECTS OF NUCLEAR
        POWER. Rev. ed. Cambridge, MA:
        MIT Press, 1975.

        A wide array of essays are included
in this volume of eight very separate and
independent articles surveying the effects
of nuclear power on society. Although
there is no cohesiveness between essays
the thought flow is manageable since they
are arranged in the order of the fuel
cycle. Essays track the uranium from the
mine through the manufacturing stages,
and then follows with the elements of
reactor operation, transportation of the
wastes, and finally the disposal of the
wastes. A few of the case studies are
quite detailed and contain many graphs and
tables. The Union of Concerned Scientists
is an advocate organization which studies
the impact of advanced technology on
society and this work will serve as a good
follow-up to a more basic text on nuclear
power.

89.   Williams, Frederick C., and David A.
        Deese, eds. NUCLEAR NONPROLIFERATION:
        THE SPENT FUEL PROBLEM. New York:
        Pergamon, 1980.

        The Spent Fuel Problem focuses on the
        by-product of the commercial nuclear
        fuel cycle and investigates what happens
        to fuel after it has been irradiated and
        withdrawn from the power reactors. One
        of the new questions which this problem
        brings forth is the issue of nuclear
        weapons proliferation resulting from
        exploited by-products. This series of
        papers discusses the problem of nuclear
        waste management as it relates to global
        security and the safeguards at nuclear
        installations. At present most spent fuel
        is being stored until the day when it can
        be economically used in fast breeder reac-
        tors--a day which may not come, according
        to the authors. Until this time, however,
        the spent fuel remains a radioactive
        nuisance. The editors provide the readers
        with some political options and venture
        some solutions to the nuclear by-product
        dilemma.

90.   Wohlstetter, Albert et al. SWORDS FROM
        PLOWSHARES: THE MILITARY POTENTIAL
        OF CIVILIAN NUCLEAR ENERGY. Chicago:
        University of Chicago Press, 1979.

        In the coming years many nations will
        be on the threshold of nuclear bomb
        production, including several countries
        which have agreed to abstain from such
        production. This book suggests that there
        is a need for a basic change in the rules
        governing the development and export of
        nuclear power, which does not necessarily
        call for a complete cessation of nuclear
        energy production. The author has made an

excellent analysis between the civil
nuclear technology which we need for
electric production and the military issue
of nuclear weaponry. This is a most factual, well-documented book.

91.   Zuckerman, Solly, Lord. NUCLEAR ILLUSION
       AND REALITY. New York: Viking, 1982.

       A British overview of the nuclear con-
troversy is presented by this one-time
Chief Scientific Advisor to the Minister
of Defense under Prime Minister Harold
Macmillan. The early U.S., UK, and USSR
involvement in the development and build-up
of nuclear warheads is traced. Lots of
"If a one-megaton bomb hit (city), x number
of people would be killed outright"
examples are provided. These are useful
if one needs to quote such destruction
projections. The illusion referred to in
the title is the idea that nuclear weapons
could ever be used as a means of defense.
He asserts that a nuclear war would be
total and mutually annihilating. Zuckerman
advocates a "no first use" policy and a
unilateral nuclear disarmament. His final
thesis is that "the existence of nuclear
weapons can neither prevent war nor
defend in war."

Neutral

92. ACCIDENT AT THREE MILE ISLAND: THE HUMAN
    DIMENSIONS. David L. Sills, C.P. Wolf,
    and Vivien B. Shelanski, eds. Boulder,
    CO: Westview, 1982.

    This is the first real sociological
    study of the impact of Three Mile Island.
    Numerous experts have analyzed organiza-
    tional behavior, the nuclear regulation
    process, the public's involvement in
    national decision making, and the conflict
    and consensus process. The stress and
    psychological effects of the TMI accident
    on the residents of the area are still
    being uncovered. Much scrutiny is given
    to nationwide community attitudes toward
    nuclear plants, nuclear responsibility,
    and the human element in the operation of
    nuclear power plants. These essays will
    prove to be landmark assessments of the
    sociological and psychological effects of
    nuclear disasters.

93.   Bolt, Bruce A.   NUCLEAR EXPLOSIONS AND
         EARTHQUAKES: THE PARTED VEIL.   New
         York: Freeman, 1976.

         The author presents an interesting
         query on a sensitive matter--how to dis-
         tinguish between earthquakes and under-
         ground nuclear tests which are forbidden
         by treaties. In very elementary terms
         Bolt provides the reader with summaries of
         the mechanics of nuclear devices and the
         elements of seismography. He also explains
         the use of nuclear devices in engineering
         and the environmental effects of under-
         ground tests. With this background the
         reader is challenged to study the earth-
         quake/nuclear test data and to decide for
         himself.

94.   Browne, Corrine, and Robert Munroe.
         TIME BOMB: UNDERSTANDING THE THREAT OF
         NUCLEAR POWER.   New York: Morrow, 1981.

         Time Bomb is a beautiful, flowing story
         that reads like a PBS documentary on the
         atomic bomb. In this superbly researched
         book the Manhattan Project, the Trinity
         Test site, and the developments of the Los
         Alamos and Oak Ridge laboratories are
         traced from their very beginnings. Playing
         a major role is J. Robert Oppenheimer whose
         involvement was as the Manhattan Project
         Supervisor and procuror of the great
         scientific minds. The distrust of nuclear
         power by Dr. John Gofman, former medical
         physicist for Lawrence Livermore Laboratory
         and now an anti-nuclear activist, is
         captured in a brief vignette about his
         early involvement in the nuclear age. This
         work is one of the most readable histories
         of the atomic age which has no emotional
         attachments to either side of the issue.

95.   THE CALIFORNIA INITIATIVE: ANALYSIS AND
      DISCUSSION OF THE ISSUES. Palo Alto,
      CA: Stanford University Institute for
      Energy Studies, 1976.

         In June 1976 Californians cast their
      votes on the California Nuclear Initiative
      (Proposition 15), a decision that was to
      determine the fate of nuclear power in that
      state. The Initiative was sure to be a
      model for other states. This book is a
      collection of six essays intended to
      provide a balanced analysis and discussion
      of the nuclear issue in California for the
      voters. No position on the Initiative is
      taken, but rather the questions and/or
      benefits of waste, jobs, risks, nuclear
      power, alternative energies, and the
      future are examined. Concise and right to
      the point, this manual can serve as a guide
      for citizen action groups as well as
      legislators contemplating the nuclear power
      issue in their state.

96.   Chicken, John C. NUCLEAR POWER HAZARD
      CONTROL POLICY. New York: Pergamon,
      1982.

         This book is more of a history of
      public outcries and movements concerned
      with the hazards of nuclear power than a
      scholarly policy book as the title suggests.
      Chicken has clearly written a general
      account of how the perceived terrors of
      nuclear warfare have masked the real, but
      different, risks of the nuclear power
      industry. We are reminded that it has
      been safer to generate power from uranium
      than from coal or oil. Indepth coverage
      is given to the counterculture as a tool
      for protest. The history of protest is
      traced to the U.S. and is a response to

Vietnam resentment and the denial of
civil rights to minorities. The movement
gained momentum in Britain where there are
still very powerful anti-nuclear lobbies
who question the need for nuclear power and
who continue to criticize the site selection
procedure for specific reactors. A careful
and thorough analysis of the risk-related
decisions and the influence which public
interest groups has had on these decisions
is provided. Briefly mentionned are the
Browns Ferry accident and the controversial
Rasmussen Report. In a finale, Chicken
clarifies for the reader England's Windscale
Inquiry of 1978.

97.   Clarke, Donald, ed. ENERGY. New York:
        Arco, 1979.

        This basic history of energy provides
a firm background in the understanding
and development of nuclear energy. From
Newton to Oppenheimer and from the theory
of energy to solar power, Clarke outlines
the essentials of power production. The
latter portion of the book covers electro-
magnetics and nuclear energy. Numerous
drawings and photographs accompany this
text designed for the general reader.

98.   Cook, Constance. NUCLEAR POWER AND LEGAL
        ADVOCACY: THE ENVIRONMENTALISTS AND THE
        COURTS. Lexington, MA: Heath, 1980.

        Here we have an attempt to record the
        actions of one type of anti-nuclear group
        and that of the legal process. Specifi-
        cally, the case of the Midland, Michigan
        nuclear power plant is followed through the
        court system. The Supreme Court did rule
        in favor of licensing and building the
        plant after several years of litigation.
        The roles of the Nuclear Regulatory Com-
        mission and the judiciary are examined
        very closely. A list of environmental
        activist groups and public interest groups
        involved in the Midland case as well as
        their activities is included.

99.   Deese, David A. NUCLEAR POWER AND RADIO-
        ACTIVE WASTE. Lexington, MA: Lexington
        Books, 1978.

        Nuclear Power and Radioactive Waste
        focuses on the possibility of sub-seabed
        disposal of radioactive waste. Deese
        discusses the advantages and disadvantages
        of this controversial method of waste
        disposal and confronts the reader with
        many economic, ethical, legal, political,
        and social issues along the way. Also
        covered is a brief history of ocean dump-
        ing and the international laws and regula-
        tions concerned. The matter of interna-
        tional public acceptance of seabed disposal
        will be the first hurdle to pass for those
        supporting this avenue for radioactive
        waste disposal.

100.   Del Tredici, Robert. THE PEOPLE OF THREE
         MILE ISLAND. New York: Sierra Club,
         dist. by Scribners, 1980.

         Interviews with thirty-seven residents
         and officials about what happened at Three
         Mile Island comprise the core of this book.
         The remarks of the people reveal their
         fears, frustrations, and outright anger
         with the utility company and its officials.
         The overriding concern is that much of the
         truth was hidden from the residents. A
         detailed chronicle of the accident is
         included. This book is very similar to
         the book by Robert Leppzer, *Voices From
         Three Mile Island* (see item number 66) and
         even includes interviews with some of the
         same people.

101.   Ebbin, Stephen, and Raphael Kasper.
         CITIZEN GROUPS AND THE NUCLEAR POWER
         CONTROVERSY. Cambridge, MA: MIT
         Press, 1974.

         This useful study is a report of citizen
         input into the nuclear power plant siting
         controversy. It was found that ignorance,
         pessimism, and doom forecasting is common
         among activist groups. Citizens groups
         cause long and costly delays in the con-
         struction and licensing of nuclear power
         plants and these costs come right back to
         the public. The authors discovered that
         lip service to public interest groups is
         rampant all the while utilities are plot-
         ting under the bargaining table to move in
         the bulldozers. The role of the regulatory
         agencies in the licensing process is
         explained as one which merely sets stan-
         dards, criteria, and rules for construc-
         tion. A history of the first contested
         licesing case of 1956 is summarized as is
         the movement to resist nuclear power in
         general.

102.   Edelhertz, Herbert, and Marilyn Walsh.
        THE WHITE-COLLAR CHALLENGE TO NUCLEAR
        SAFEGUARDS. Lexington, MA: Heath, 1978.

        We have all been informed about the
        activists and terrorists who pose threats
        to nuclear safety but here the authors
        provide us with a relatively unexposed
        culprit--employees of nuclear installations
        who break the security systems for a price.
        Employee sabotage and the diversion of
        classified materials presents special prob-
        lems in nuclear security and safety. The
        authors point out that this area of sabo-
        tage is too often overlooked as the regu-
        latory agencies are primarily concerned
        with the external threats. As the nuclear
        materials market expands the diversion
        and sales of illicit materials becomes more
        profitable. *White-Collar Challenge*
        recommends a study of white collar crime
        in the nuclear industry and the formulation
        of safeguards designed to deter white
        collar crime in the atomic industry. The
        authors treat this topic as if this is a
        potential crime whereas in actuality such
        criminal activities have already been
        documented. Readers interested in this
        rarely written about topic will find this
        a stimulating book.

103.   Fenn, Scott. THE NUCLEAR POWER DEBATE:
        ISSUES AND CHOICES. New York: Praeger,
        1981.

        The nuclear industry continues to be
        plagued by economic, political, and tech-
        nological uncertainties. Many industry
        trends are brought to light including the
        observation that utility companies have
        begun a self-imposed moratorium on nuclear
        plant orders. Time is needed to plan and

study nuclear plant performance data. The
Nuclear Regulatory Commission, a once
strong advocate of nuclear power, has even
made decisions that have had an adverse
impact on the economic and political
acceptance of nuclear power. Regardless
of the neutrality of this book, the author
agrees that the international development
of nuclear power will aid in the proliferation of nuclear weapons.

104.   Fermi, Laura. ATOMS FOR THE WORLD: UNITED
         STATES PARTICIPATION IN THE CONFERENCE
         ON THE PEACEFUL USES OF ATOMIC ENERGY.
         Chicago: University of Chicago Press,
         1957.

         The author is the widow of Enrico Fermi
         and *Atoms for the World* is her story of
         the U.S. role at the 1955 International
         Conference on the Peaceful Uses of Atomic
         Energy held in Geneva, Switzerland. The
         book does not pretend to be the official
         recorded history of the conference but
         rather it is only one layperson's view and
         impressions of the proceedings. Covered
         in the summary are the Russian scientists
         in Geneva, the technical exhibits which she
         found to be fascinating, power, radiation
         hazards, and much more. This is an easy
         to comprehend encapsulation of the first
         conference of this type. Although this is
         not a recently published book, the signi-
         ficance of this conference and Laura
         Fermi's comments are historically signifi-
         cant.

105.   Francis, John, and Paul Abrecht, eds.
        FACING UP TO NUCLEAR POWER: RISKS AND
        POTENTIALITIES OF THE LARGE-SCALE USE
        OF NUCLEAR ENERGY. Philadelphia:
        Westminster, 1976.

        Written by leading experts this collec-
        tion of essays discusses the economic,
        ecological, and ethical issues of nuclear
        power. Predictions are made as to what we
        can expect by the year 2001 by way of
        nuclear energy. Included in this book is
        the renowned and controversial report of
        the "Ecumenical Hearing on Nuclear Energy"
        sponsored by the World Council of Churches
        held in Sweden in 1976. As is evident from
        the tone of many of the essays, the real
        debate over the future of nuclear energy
        has scarcely begun.

106.   Gompert, David C. et al. NUCLEAR WEAPONS
        AND WORLD POLITICS: ALTERNATIVES FOR
        THE FUTURE. New York: McGraw-Hill,
        1977.

        Gompert as editor for this book arranged
        for four other writers to think through
        the situation of what nuclear weapons will
        pose for international peace and progress
        through the 1980s. Each approached the
        same set of circumstances and presented
        very different responses and conclusions
        in a thought provoking manner. The author,
        in a concluding essay, wraps everything up
        and places the other four essays into
        perspective. Due to the caliber of the
        writing and the serious consideration
        required to understand the greater question,
        this is a book for the advanced nuclear
        reader.

107.   Halacy, Dan. NUCLEAR ENERGY. New York:
        Franklin Watts, 1978.

        Directed toward the junior high reader,
        this is an outstanding starter book on
        nuclear energy. Anyone unfamiliar with
        nuclear power will have a basic understand-
        ing of things nuclear after reading this
        book. Contents include brief discussions
        of atomic power, nuclear fuel, how a
        nuclear power plant operates, the dangers
        of nuclear energy, and a chronology of
        nuclear energy development. A useful
        reading guide for further study follows
        the text.

108.   Hellmann, Richard, and Caroline J.C.
        Hellmann. THE COMPETITIVE ECONOMICS
        OF NUCLEAR AND COAL POWER. Lexington,
        MA: Lexington Books, 1983.

        The authors, both economists, analyze
        the economics of nuclear and coal power
        plants which have been newly ordered and
        are expected to begin operation between
        1985 and 1995. The problem of underesti-
        mating construction costs and cost over-
        runs is a dilemma shared by the architect,
        the engineers, the manufacturers, and the
        public. Although this is a financial study
        of the industry, and therefore economically
        technical, general readers will find it
        useful for a look at how management views
        the cost problem and for comparing the
        nuclear vs. coal power question. Many
        charts, graphs, and tables accompany the
        text. The book covers the design and
        technological risks of nuclear power, the
        human factors, waste disposal, the decom-

missioning process, and several case
studies like the AEC, ERDA, NRC, and Exxon
cost studies.

109. International Scientific Forum on an
Acceptable Nuclear Energy Future of the
World. NUCLEAR ENERGY AND ALTERNATIVES.
Cambridge, MA: Ballinger, 1978.

This work comprises the proceedings of a
conference held from November 7-11, 1977
at the Center for Theoretical Studies of
the University of Miami. Nearly fifty
educators, scientists, industrialists, and
governmental leaders participated by
presenting papers at this international
forum on nuclear energy. Some of the
papers may be considered highly technical
but there are plenty which the well-
informed reader can comprehend. In this
latter category is a series of papers
devoted to a worldwide energy inventory,
health and safety matters, the nuclear
proliferation controversy, and solar energy
as an alternative. The conference may not
go down in history as a landmark symposium
but, nevertheless, many of the papers
presented are noteworthy and somewhat controversial. This balanced collection does
contribute to a better understanding of
the greater nuclear power question.

110.   IS NUCLEAR POWER SAFE? Washington, DC:
         American Enterprise Institute, 1975.

         This small book is the edited transcript
         of an American Enterprise Institute Round
         Table discussion which was moderated by
         Melvin Laird, former Secretary of the
         Defense. The five participants were Ralph
         Nader, consumer advocate; Ralph Lapp,
         physicist; Craig Hosmer, former California
         Congressman; and Lawrence Moss, nuclear
         engineer and former Sierra Club president.
         This mix of people in the same room pre-
         sents a potential volatile situation but
         they do manage a most civil discourse on
         the nuclear safety issue. Although this
         is brief, useful and potent arguments are
         given for and against the continued use of
         nuclear power.

111.   Kaku, Michio, and Jennifer Trainer, eds.
         NUCLEAR POWER: BOTH SIDES: THE BEST
         ARGUMENTS FOR AND AGAINST THE MOST
         CONTROVERSIAL TECHNOLOGY. New York:
         Norton, 1982.

         This anthology provides a well-balanced
         analysis of the nuclear power debate
         through it twenty-five contributed essays.
         Most of the writers are vocal proponents
         or critics of the nuclear industry and
         each is probably represented elsewhere in
         this bibliography with his or her own book.
         This would be another excellent first book
         to read on the nuclear power controversy.

112. Kuhn, James W. SCIENTIFIC AND MANAGERIAL
      POWER IN THE NUCLEAR INDUSTRY. New
      York: Columbia University Press, 1966.
      Foreword by Eli Ginzberg.

      Here is a work which examines the extent
      to which manpower influenced the develop-
      ment of the new technology pre-World War II.
      Most of these gifted scientists (refugees
      from Hitler and Mussolini) were responsible
      for the harnessing of nuclear energy in
      America. In fact, the atomic bomb itself
      and the successful conclusion to World
      War II can be attributed to this talented
      group of foreign-born scientists. In later
      years the development of peaceful nuclear
      power became an interdisciplinary project
      involving a diverse group of scientists
      and mathematicians. Professor Kuhn also
      discusses the role and errors that American
      corporations made when hiring engineers to
      direct the company research in nuclear
      power. He surmises that management lacked
      the necessary technical competence to
      understand their role and the associated
      risks, so they merely handed over the
      decision making powers to the technical
      staffs. In so doing they lost touch and
      often did not understand what they were
      getting into until it was too late, and
      perhaps too costly to back out. Also
      covered is the critical role that Admiral
      Hyman Rickover played in the development
      of nuclear power, the problems facing the
      utility companies, and the skill obso-
      lescence that occurs early in research and
      development groups. This work is a
      significant historical document of the
      decision making powers behind the present
      nuclear industry.

113.   Lewis, Richard S. THE NUCLEAR POWER
        REBELLION: CITIZENS VS. THE ATOMIC
        INDUSTRIAL ESTABLISHMENT. New York:
        Viking, 1972.

        In a narrative format Richard Lewis
        covers numerous events in the young history
        of atomic energy. Most of the incidents
        detail the involvement of the "down on the
        farm" citizenry and their reactions to a
        nuclear power plant in their backyard.
        Many court cases are highlighted for easy
        interpretation. Interesting commentary is
        provided on radiation hazards and on the
        move to build breeder reactors.

114.   Martin, Laurence, ed. STRATEGIC THOUGHT
        IN THE NUCLEAR AGE. Baltimore: Johns
        Hopkins University, 1980.

        Some interesting and challenging argu-
        ments are presented in the seven contri-
        buted essays which comprise *Strategic
        Thought*. As a whole the essays record
        strategic thought since 1945 on a global
        scale and in light of a growing nuclear
        arsenal. Topics covered include the util-
        ity of the military in world politics, the
        economics of national security, threat as
        perceived in the intelligence process,
        limited war from minimal to nuclear, the
        history of strategic forces and targeting,
        managing international crises, and arms
        control. All authors concur that military
        strength is a fact of life and that arms
        will not go away. A concern is expressed
        that governments learn to manage via arms
        control, not the elimination of arms which
        could probably never be achieved in most of
        our lifetimes. Numerous methods for
        increased arms control and the many bar-
        riers which exist are analyzed. The ideas
        presented in this book are not necessarily

new and innovative but they are presented
in a readable, professional manner.

115.   Metzger, H. Peter. THE ATOMIC ESTABLISH-
        MENT. New York: Simon & Schuster,
        1972.

        This is a historically valuable work in
that the author shows how the U.S. Congress
and the Atomic Energy Commission changed
from enemies to friends. Indepth discussions are provided on nuclear weapons and
radioactive waste with several solutions
for their management being offered.
Metzger suggests that the problems we have
today with nuclear energy are not technical
ones, but rather they are politically
rooted. This is a hefty volume of easy
reading for those with a nuclear interest.

116.   Moss, Thomas H., and David L. Sills.
        THE THREE MILE ISLAND NUCLEAR ACCIDENT:
        LESSONS AND IMPLICATIONS. New York:
        New York Academy of Sciences, 1981.
        (Annals of the New York Academy of
        Sciences, v. 365)

        In 1980 the New York Academy of Sciences
held a conference on the nuclear incident
at Three Mile Island. The keynote theme
was that "the central problem of nuclear
energy is not waste disposal or even the
connection between proliferation and
nuclear power. It is, rather, reactor
safety", according to Alvin Weinberg. The
conference was attended by many of the same
people who were touched by the accident
just one year earlier and who were now

present to give their personal views.
Discussions centered on the technical
history of the accident, the predicted
likelihood of such an event, the radiation
released, the problem of decontaminating
the site, the behavior of individuals and
organizations involved, and the attitudes
of the local residents. A most interesting
segment deals with the relationship between
the nuclear authorities and the press. To
the anti-nuclear people TMI showed that the
industry operates on the brink of disaster.
To the nuclear proponents it merely con-
firmed the safety of the system. One
question which was asked at the conference
was whether we should reconsider the gas-
cooled reactor, one which depends less on
engineered safety features and more on
inherent characteristics to prevent accidents.

117.   Nero, Anthony V. A GUIDEBOOK TO NUCLEAR
        REACTORS. Berkeley: University of
        California Press, 1979.

        Informative, rather than argumentative,
this book explains the major features of
reactors and the issues surrounding their
use. The book begins with an outline of
the main features of a nuclear power plant
and explains the effect which reactors
have on their environment. Many pages are
devoted to discussions of pressurized water
reactors, boiling water reactors, heavy
water reactors, and gas-cooled thermal
reactors. Nero also considers the sources
of uranium and the various nuclear fuel
cycles. Under the premise that nuclear
power demands will accelerate at a slow
pace in the U.S., the author provides
figures which should help to decide how
much nuclear power is needed, in what form,

and which type of reactor is best suited to
that need. The focus is on reactor systems
found in, or of interest to the United
States. This means that the many British,
French, and Russian reactor types are miss-
ing from this guidebook. Dr. Nero has provided the non-scientist with a concise
source of information on a variety of
nuclear systems all the while relating them
to the controversial issues of safety,
waste, and weapons.

118.   NUCLEAR POWER IN AMERICAN THOUGHT.
       Washington, DC: Edison Electric Insti-
       tute, 1980. (Decisionmakers Bookshelf
       Series, v. 8)

       In an approach very different from
other books in this bibliography, *Nuclear
Power in American Thought* handles the
nuclear power debate with non-technical
arguments involving ethics, history, philosophy, politics, and psychology. The book
is a collection of four essays which attempt
to show how these humanistic concerns affect
and will continue to affect American thought
on the nuclear issue. Contributors are
Dr. Andrew Hacker, a professor of political
science; Dr. Robert L. DuPont, psychiatrist
and president of the non-profit Institute
for Behavior and Health, Inc.; Dr. William
Barrett, a professor of philosophy; and Dr.
Margaret Maxey, Assistant Director of the
South Carolina Energy Research Institute.
With such an intellectual group of authors
one might expect a work that is untouchable
by most general readers but such is not the
case. The essay by Dr. DuPont on nuclear
phobia is outstanding as are William Barrett's
predictions on the American acceptance of
nuclear power.

119. Oak Ridge Associated Universities. Institute
      for Energy Analysis. ECONOMIC AND
      ENVIRONMENTAL IMPACTS OF A U.S. MORA-
      TORIUM, 1985-2010. Cambridge, MA:
      MIT, 1979.

          Numerous futurists contributed to the
      research and conclusions in this book which
      examines the impact of a nuclear freeze.
      Five distinct economic implications are
      observed: future cost of electricity,
      regional dislocations, the nuclear industry
      impacts, the effect on the coal industry,
      and the international impacts. Some of the
      conclusions derived include the prediction
      that energy demand will increase at a slow
      rate and that nuclear power is cheaper than
      fossil-generated power in most areas of the
      U.S. It is also understood that a nuclear
      moratorium will require that more coal be
      mined and the burning of this coal will
      increase emissions into the air, perhaps
      changing the worldwide climate. Dozens of
      tables, charts, and figures are provided to
      help you draw your own conclusions. How-
      ever, all of these studies and predictions
      cannot answer the question, what will be
      the energy supply and demand beyond the year
      2010?

120. Patterson, Walter C. NUCLEAR POWER.
      Baltimore, MD: Penguin, 1976.

          This small paperback may be used as a
      starting point on the path to developing a
      nuclear awareness. In very simple terms
      the elements of the nuclear predicament are
      explained as they relate to history, costs,
      reactors, and the future. An attempt is
      made to answer many of the pressing

questions and an ample reading list is
provided for further clarification.

121. Peat, David F. THE NUCLEAR BOOK. Ottawa,
     Canada: Deneau and Greenberg, 1979.

     Canadian scientist and journalist
David Peat has written this book as a guide
to nuclear safety in Canada. The Three Mile
Island incident prompted Canadians to look
more closely at their CANDU (Canadian-Deu-
terium-Uranium) reactor, a heavy-water
reactor considered to be safe and reliable.
The CANDU reactor is compared to other types
of reactors worldwide in terms of safety,
economics, energy demands, and the technology which is available. Peat explains why,
in his opinion, the CANDU reactor was the
best choice for Canada at the time. This
work also provides elementary descriptions
of nuclear reactors, waste, risk, nuclear
industry regulations, alternatives for the
future, and many more critical issues.

122. Pringle, Laurence. NUCLEAR POWER: FROM
     PHYSICS TO POLITICS. New York:
     Macmillan, 1979.

     The development of nuclear power is
traced from Hiroshima and Nagasaki to the
present day with predictions for the year
2000. Readers are informed of the secrecy
surrounding the development of atomic power
and explore decades of governmental confu-
sion as to who promotes and who regulates
nuclear power. With this as a background,
Pringle looks at the controversy itself and
presents both sides of the question to
readers. This is a very easy-to-read high
school level book with many photographs.

123.   Shrader-Frechette, K.S.   NUCLEAR POWER
         AND PUBLIC POLICY: THE SOCIAL AND
         ETHICAL PROBLEMS OF FISSION TECHNOLOGY.
         Dordrecht, Holland: D. Reidel, 1980.

         A variety of problems are presented as
         the author examines the social, political,
         and ethical issues which have heretofore
         been ignored by the government when making
         nuclear power assessments. The author
         argues that the current policy regarding
         nuclear technology is unjust, that to
         generate long-lived nuclear wastes is
         reprehensible, and that a core meltdown
         is improbable on both logical and
         scientific grounds. Very brief coverage
         is given to the economics of nuclear power
         and nuclear safety.

124.   Stephens, Mark.   THREE MILE ISLAND.   New
         York: Random House, 1981.

         This is essentially a detailed chronicle
         of what happened during the last week of
         March 1979 at Three Mile Island. In a
         very matter of fact style the author
         relates the occurrences going on inside
         the power plant and how the Nuclear Regu-
         latory Commission and other public
         officials responded. Some commentary on
         how the press covered the affair and their
         antagonistic relationship with the offi-
         cials is included. The author's conclusion
         is that TMI officials and the state and
         national regulatory agencies considered
         the chance of an accident so remote that
         they were really unprepared to deal with
         the incident or to even protect the
         people.

125.   TOSCA: THE TOTAL SOCIAL COST OF COAL AND
       NUCLEAR POWER. Linda Gaines, R. Stephen
       Berry, and Thomas V. Long, eds.
       Cambridge, MA: Ballinger, 1979.

       Social costs are defined as the direct
       cost of generating power combined with the
       indirect costs of human health, the
       environment, waste, etc. *TOSCA* includes a
       model for estimating these social costs of
       coal and nuclear power generation to the
       year 2004. The authors discuss research
       and development expenditures and venture a
       controversial claim that pressurized water
       and boiling reactors are identical. This
       work is primarily concerned with the econo-
       mic factors of power generation and the
       discussion of the model which the authors
       have developed. The value of the book lies
       in its use as an energy reference and for
       those hard-to-find cost comparisons.

126.   Webb, Richard E. THE ACCIDENT HAZARDS OF
       NUCLEAR POWER PLANTS. Amherst, MA:
       University of Massachusetts Press, 1976.

       Four classes of reactor accidents in
       which the fuel could overheat and other
       accidents which may be caused by design and
       error are graphically presented. The topic
       of nuclear sabotage is briefly discussed as
       is the controversial Rasmussen Report, a
       congressional reactor safety study. The
       author stresses the value of nuclear power
       as an alternative fuel but cautions that we
       should assess the hazards first to deter-
       mine whether or not the risks are accept-
       able. The negative aspects of nuclear power
       may appear to be the stronger element but
       Webb does present an equally balanced,
       unbiased report of nuclear power.

127.   Weiss, Ann E. THE NUCLEAR QUESTION.
          New York: Harcourt Brace Jovanovich,
          1981.

          Aimed at the high school aged or the
          uninformed reader, *The Nuclear Question*
          summarizes the story of nuclear power from
          its promise, development, costs, benefits,
          and dangers. Briefly covered are some of
          the events which led to the present day
          debate over the use of nuclear power. An
          even balance of arguments for and against
          nuclear power are provided. Numerous
          questions about the future of nuclear power
          and other sources of energy complete this
          introduction to the nuclear controversy.

128.   United States. President's Commission on
          the Accident at Three Mile Island.
          THE NEED FOR CHANGE, THE LEGACY OF TMI:
          REPORT OF THE PRESIDENT'S COMMISSION ON
          THE ACCIDENT AT THREE MILE ISLAND.
          Washington, DC: U.S. Government Printing
          Office, 1979.

          Two weeks after the TMI incident Presi-
          dent Carter appointed a twelve member com-
          mission to complete a comprehensive study
          of the event. Dr. John G. Kemeny, Presi-
          dent of Dartmouth College, chaired the
          Commission which was charged with making a
          technical assessment of the events and
          their causes, to analyze the management, to
          evaluate the response of the Nuclear Regu-
          latory Commission and other emergency pre-
          paredness authorities, to assess the flow
          of information concerning the events at TMI,
          and to make appropriate recommendations.
          The findings and the recommendations of the
          Commission are too numerous to list here
          but the general tone is one of suspicion
          and concern over the practices of the NRC.
          Numerous photos, an account of the accident,

and a staff list is appended to the
report.

129.   Winter, John, and David Conner. POWER
        PLANT SITING. New York: Van Nostrand
        Reinhold, 1978.

        Although primarily geared to the archi-
        tects and engineers who are concerned with
        the siting process, this book is valuable
        to others as a guide to understanding the
        care, analysis, and study which goes along
        with site selection. The viewpoints of the
        public, the utility companies, and the
        governmental regulatory agencies are con-
        sidered. The book is thorough in its
        discussions of the energy crisis, legisla-
        tion, land and sea siting possibilities,
        nuclear concerns, power plant design, and
        future considerations. Many charts as well
        as a sample environmental impact statement
        accompany the text.

Author Index

Subject Index

Accidents 4,5,38,46,49,
54,55,62,75,126,128
Alternative Energy 12,
24,27,33,36,42,45,57,
60,67,83,95,108,109,
127
Arms Control. *See under*
Weapons Proliferation.
Browns Ferry 42,49,96
Court Cases 81,98,101,
113
Environment 4,7,11,88,
105,119
Fast-Breeder Reactors. *See*
under Reactors, Fast-
Breeder.
Fission Reactors. *See*
under Reactors,
Fission.
Foreign Policy 13,17,43,
47,51,62,69,77,79,90,
106
Fusion Reactors. *See*
under Reactors, Fusion.
Gas-Cooled Reactors. *See*
under Reactors, Gas-
Cooled.
Heavy-Water Reactors. *See*
under Reactors, Heavy-
Water.
Light-Water Reactors. *See*
under Reactors, Light-
Water.

Moratorium 42,47,52,
64,84,103,119
Nuclear Energy, benefits
of 8,22,25,28
Nuclear History 2,6,8,
14,20,21,22,23,26,30,
31,38,41,45,53,60,63,
70,86,94,96,97,104,
107,112,114,115,118,
120,122,127
Nuplexes 10,28
Oppenheimer, Robert J.
20,31,63,72,76,94,97
Price-Anderson Act 57
Radiation 39,49,58,73,
76,82,84,113
Rasmussen Report 96,126
Reactors, Fast-Breeder
4,7,16,29,55,83,113,
117,121
Reactors, Fission 4,6,
10,19,72,73,117,121,
123
Reactors, Fusion 2,6,
10,117,121
Reactors, Gas-Cooled 4,
30,117
Reactors, Heavy-Water 4,
30,117,121
Reactors, Light-Water 3,
4,30,117,121
Reactors, Thermal 7,117

Title Index